Global Issues Series

General Editor: **Jim Whitman**

This exciting new series encompasses three principal themes: the interaction of human and natural systems; cooperation and conflict; and the enactment of values. The series as a whole places an emphasis on the examination of complex systems and causal relations in political decision-making; problems of knowledge; authority, control and accountability in issues of scale; and the reconciliation of conflicting values and competing claims. Throughout the series the concentration is on an integration of existing disciplines towards the clarification of political possibility as well as impending crises.

Titles include:

Brendan Gleeson and Nicholas Low (*editors*)
GOVERNING FOR THE ENVIRONMENT
Global Problems, Ethics and Democracy

Roger Jeffery and Bhaskar Vira (*editors*)
CONFLICT AND COOPERATION IN PARTICIPATORY NATURAL RESOURCE MANAGEMENT

Ho-Won Jeong (*editor*)
GLOBAL ENVIRONMENTAL POLICIES
Institutions and Procedures

W. Andy Knight
A CHANGING UNITED NATIONS
Multilateral Evolution and the Quest for Global Governance

W. Andy Knight (*editor*)
ADAPTING THE UNITED NATIONS TO A POSTMODERN ERA
Lessons Learned

Graham S. Pearson
THE UNSCOM SAGA
Chemical and Biological Weapons Non-Proliferation

Andrew T. Price-Smith (*editor*)
PLAGUES AND POLITICS
Infectious Disease and International Policy

Michael Pugh (*editor*)
REGENERATION OF WAR-TORN SOCIETIES

Bhaskar Vira and Roger Jeffery (*editors*)
ANALYTICAL ISSUES IN PARTICIPATORY NATURAL RESOURCE MANAGEMENT

Simon M. Whitby
BIOLOGICAL WARFARE AGAINST CROPS

Global Issues Series
Series Standing Order ISBN 0–333–79483–4
(*outside North America only*)

You can receive future titles in this series as they are published by placing a standing order. Please contact your bookseller or, in case of difficulty, write to us at the address below with your name and address, the title of the series and the ISBN quoted above.

Customer Services Department, Macmillan Distribution Ltd, Houndmills, Basingstoke, Hampshire RG21 6XS, England

Biological Warfare against Crops

Simon M. Whitby
Research Fellow
Department of Peace Studies
University of Bradford

palgrave

First published 2002 by
PALGRAVE
Houndmills, Basingstoke, Hampshire RG21 6XS and
175 Fifth Avenue, New York, N.Y. 10010
Companies and representatives throughout the world

PALGRAVE is the new global academic imprint of
St. Martin's Press LLC Scholarly and Reference Division and
Palgrave Publishers Ltd (formerly Macmillan Press Ltd).

ISBN 0–333–92085–6

This book is printed on paper suitable for recycling and made from fully managed and sustained forest sources.

A catalogue record for this book is available from the British Library.

Library of Congress Cataloging-in-Publication Data
Whitby, Simon M., 1960–
Biological warfare against crops/Simon M. Whitby.
p. cm.—(Global issues)
Includes bibliographical references (p.).
ISBN 0–333–92085–6 (cloth)
1. Biological warfare. 2. Plant diseases—Epidemics.
3. Agriculture—Defense measures. I. Title. II. Series.

UG447.8.W49 2001
358'.38—dc21

2001032130

10 9 8 7 6 5 4 3 2 1
11 10 09 08 07 06 05 04 03 02

Printed and bound in Great Britain by
Antony Rowe Ltd, Chippenham, Wiltshire

Contents

List of Figures

List of Tables

Preface

Germ warfare or biological warfare (BW) can be described as the intentional cultivation or production of pathogenic bacteria, fungi, viruses, and their toxic products, as well as certain chemical compounds, for the purpose of deliberately producing disease or death. Portrayed in such terms the horror of biological weapons is usually thought of as the intentional exposure of a human population to deadly diseases, such as anthrax or plague.

In contrast to growing concerns about biological warfare aimed more directly at humans, biological warfare against crops has received little attention. This form of warfare involves the deliberate initiation of crop disease epidemics through the application to crops of plant pathogens such as bacteria, parasitic fungi, and viruses. When weaponised, such pathogens can be used to cause disease in plants and the large-scale destruction of food and cash crops.

This book argues that biological warfare against crops is a form of warfare that could have potentially devastating consequences and may prove to have considerable future potential. It shows that investigations into anti-crop warfare have formed a significant component of all known offensive biological warfare programmes and argues that, where states are intent on acquiring a capability to wage biological warfare, such activities are likely to include investigations into acquiring a capability to wage biological warfare against crops.

This book is the first substantive account of this subject. The public record on anti-crop biological warfare remains highly variable, with most of the available information relating to the programme in the United States during the Second World War and the early Cold War years. The book explores the subject in some depth by attempting to place what is known about the US programme in the context of far less complete information on Axis and Allied biological warfare programmes between the Second World War and the late 1950s, as well as the more recent Iraqi and Soviet programmes. The account of Anglo-American activities after the Second World War is of particular value, not least in relation to the original source material obtained from official British and American sources, the latter providing a valuable insight into the nature of transatlantic collaboration in this field.

The book is of direct policy relevance. It places concerns about proliferation in the context of the initiative to strengthen the international legal prohibition against this form of warfare. Finally, it argues that related control regimes on the safe handling, storage, use, and transfer of pathogens that pose a risk to human health and the environment are of direct relevance to the initiative to strengthen the Biological Weapons Convention. If effectively and efficiently implemented, such regimes will be mutually reinforcing and the risks posed from outbreaks of disease in plants, however caused, will be minimised.

Acknowledgements

I would like to express my sincerest gratitude to those who have assisted in the preparation on this book. Given the dearth of literature associated with the subject matter it would simply not have been possible to compile a book on this topic had it not been for the painstaking and studious work of those at Sussex Harvard Information Bank located at the Science Policy Research Unit, University of Sussex. Therefore, thanks must go to Julian P. Perry Robinson who has worked tirelessly over many years to further the public understanding on issues related to the prohibition and control of chemical and biological warfare, and who on many occasions, has generously shared his facilities and expertise with me. Thanks must also go to the numerous contributors to this invaluable collection of literature.

A considerable debt of gratitude is owed to my colleague Professor Malcolm R. Dando. Thanks also to Professor Paul F. Rogers, who along with Malcolm Dando gave me the opportunity to conduct the research for this book.

Thanks must also go to my friends and colleagues in North America. In particular Milton Leitenberg has provided invaluable insights into the subject matter. He has also been of assistance in providing useful documentation and references. Thanks go to Dan Plesch and the British American Security Information Council (BASIC). Thanks also go to Dr Leo L. Laughlin.

Additionally, I am thankful for the help of the following: Dr Alastair Hay, Department of Chemical Pathology, University of Leeds; Professor Graham S. Pearson CB, University of Bradford; Jez Littlewood, University of Bradford; David Yates (deceased), formerly University of Bradford; Brian Balmer, University College, London; Jean Pascal Zanders, Stockholm International Peace Research Institute (SIPRI), Sweden; Henrietta Wilson (formerly) Harvard Sussex Program; Anne Costigan, Life Science Librarian, and Inter-Library Loans, J.B. Priestley Library, University of Bradford; Ewen Buchanan and Steve Black, United Nations Special Commission (UNSCOM) on Iraq. Special thanks also go to Mandy Oliver.

Apologies for those that have been unintentionally omitted from the above acknowledgements. I alone take full responsibility for any mistakes or inaccuracies that appear in the pages that follow.

List of Abbreviations

AAF	American Air Force
BTWC	Biological and Toxin Weapons Convention
BW	Biological Warfare
CBD	Convention on Biodiversity
CEB	Centre d'Etudes du Bouchet
CEBAR	Chemical, Biological and Radiological
CW	Chemical Weapons
CWS	Chemical Warfare Service
DRPC	Defence Research Policy Committee
FAO	Food and Agriculture Organization
FSA	Federal Security Agency
FY	Fiscal Year
IAEA	International Atomic Energy Authority
IPPC	International Plant Protection Convention
JRDB	Joint Research and Development Board
NAS	National Academy of Sciences
NDRC	National Defense Research Council
NPT	Non Proliferation Treaty
NRC	National Research Council
OSRD	Organization for Scientific Research and Development
OTA	Office of Technology Assessment
SAC	Supreme Allied Command
SIPRI	Stockholm International Peace Research Institute
UNEP	United Nations Environmental Program
UNSCOM	United Nations Special Commission (on Iraq)
USAF	United States Air Force
VEREX	Verification Experts
WBC	War Bureau of Consultants
WMD	Weapons of Mass Destruction
WRS	War Research Service
WSEG	Weapons Systems Evaluation Group

Glossary

Anthrax An infectious disease of humans and animals caused by *Bacillus anthracis*.

Autoclaving Autoclaving refers to a process of sterilisation. Autoclaving takes place in a stainless steel vessel that is designed to withstand the steam pressures employed in sterilisation.

Botulinum toxin Toxins caused by *Clostridium botulinum*. The cause of a form of food poisoning.

Brucellosis A disease of humans and animals caused in humans by *Brucella* bacteria. The human form of the disease is known as undulant fever.

Colorado beetles Beetles native to North America that pose a serious threat to potato crops.

Dysentery Can be caused by *Shigella* bacteria. Dysentery can cause painful inflammation and infection of the intestinal tract.

Epidemic Severe widespread outbreak of disease.

Epiphytotic Plant disease epidemic.

Foot and Mouth Disease (FMD) Disease of livestock including cattle, sheep and pigs. Caused by FMD virus.

Glanders Disease of mammals (mainly of the horse family) but can affect humans. Caused by the bacteria *Pseudomonas mallei*.

Helminthosporium oryzae Causal agent of brown spot of rice.

Hypertophy Overgrowth in plants due to the abnormal enlargement of cells.

Hyperplasia Overgrowth in plants due to increase in cell division.

Hypha A single part of mycelium.

Mycelium The mass containing hypha that makes up the body of fungus.

Mycoplasmas Microorganisms that have no cell wall.

Necrotic Death and/or discoloration of plant tissue due to disease.

Pathogen A disease producing entity or organism.

Phytopathogenic A term applied to an entity or organism that is disease producing.

Phytophthora infestans The causal agent of late blight of potatoes.

Piricularia oryzae Causal agent of rice blast.

Plague A severe infectious disease caused by the bacteria *Yersinia pestis*. The mode of dissemination of the disease in humans is often via infected fleas. In pulmonary plague infection can occur through inhalation of bacteria.

Puccinia graminis Causal agent of wheat rust.

Rinderpest An acute disease of livestock caused by a highly pathogenic virus.

Smallpox An acute communicable disease in humans caused by a virus.

Sporangium An entity containing asexual spores, or in some cases a single spore.

Spore The reproductive entity of fungi. Can contain one or more cells.

Typhoid A disease of humans and animals caused in humans by *Salmonella typhi*.

1
Introduction

BW is the deliberate spreading of disease among humans, animals, and crops. This book examines deliberate disease against crops as a form of biological warfare.

At the time of the Gulf War in 1991, it was widely recognised that Iraq already had a chemical weapons capability, having used such weapons in the war with Iran in the 1980s and also against elements of its own Kurdish population, notably at Hallebjah[1] in March 1988. There was also some concern that Iraq had a BW programme, but there was little information in the public domain at the time of the 1991 war concerning the extent of this programme. It was not known for example, whether biological weapons were deployed during the war, and whether the Iraqi BW programme was concerned with BW systems intended for use against humans, animals or both. In particular, there did not appear to be any evidence that Iraq had been interested in anti-crop weapons.

It is now known that the Iraqi BW programme included a specific component concerned with the development of anti-crop agents, principally fungal pathogens, which might be used deliberately to cause epidemics in grain crops of enemy states. This was part of a comprehensive BW programme which included, by 1991, the weaponisation of anti-personnel BW systems and their deployment for use.

The example of the Iraqi anti-crop warfare programme is possibly the sole independently verified example of published data on this subject for any country in the past two decades, and may have considerable implications for future developments in biological warfare and its control.

The aim of this book is to show that where states are intent on acquiring a capability to wage biological warfare and have access to modern biology such activities are likely to include investigations into biological warfare against crops. In order to assert this claim this book will first, examine Iraq's attempts to acquire a capability to wage biological warfare; and secondly, it will examine the biological warfare programmes that emerged during the course of the middle years of the twentieth century. While a full description of the biological warfare programmes under review would require a detailed account of both defensive and offensive preparations, for simplicity, this book will concentrate on investigations into offensive aspects of such programmes, that is, on investigations into the development of weapons. The focus throughout will be on investigations relating to deliberate disease against crops.

However, it is a requirement of this book to consider developments in anti-crop biological warfare in respective states alongside anti-personnel and anti-animal BW preparations. Moreover, investigations into anti-crop biological warfare were inextricably linked to investigations conducted into anti-crop warfare with chemical agents, and where it has been necessary to do so, developments relating to investigations into both chemical and biological anti-crop warfare will be considered together.

The acquisition of a militarily significant capability to wage biological warfare is a complex multi-step process, and an offensive biological warfare programme will have a number of essential characteristics. In this connection a Biological Weapon Acquisition Table (Table 1.0) produced in a study conducted by the US Office of Technology Assessment (OTA)[2] in 1993 shows some of the necessary steps that are required in obtaining an operational capability in offensive biological warfare.

The OTA study identified eight steps[3] in the biological weapons acquisition process. Table 1.1 shows what a militarily significant biological warfare capability would involve, according to the OTA.

In accordance with the OTA's biological weapons acquisition process, a state intent on acquiring an operational military capability to wage anti-crop biological warfare would therefore be required to organise its activities in order that it successfully negotiates each step in the acquisition process outlined above. It is important to note, however, that not all activities addressed in this book resulted in an offensive capability to wage anti-crop biological warfare. Where such activities did not result in an offensive capability to wage anti-crop BW, the book will show that offensive anti-crop BW research still represented a significant component in all known BW programmes, and

Table 1.1 Key steps in the acquisition of a militarily significant BW capability

1. Establishment of one or more facilities and associated personnel with organisational and physical provisions for the conduct of work
2. Research on microbial pathogens and toxins, including the isolation or procurement of virulent or drug-resistant strains
3. Pilot production of small quantities of agent in flasks or small fermenter systems
4. Characterisation and military assessment of the agent, including its stability, infectivity, course of infection, dosage, and the feasibility of aerosol dissemination
5. Research, design, development, and testing of munitions and/or other dissemination equipment
6. Scaled-up production of agent (possibly in several stages) and freeze–drying
7. Stabilisation of agent (e.g. through micro-encapsulation) and loading into spray tanks, munitions, or other delivery systems, and
8. Stockpiling of filled or unfilled munitions and delivery vehicles, possibly accompanied by troop training, exercises, and doctrinal development.

on that basis, I will argue that it is a reasonable assumption that offensive anti-crop biological warfare research may represent a significant component in (1) current programmes about which little public information is available, and (2) future biological warfare programmes.

Background

BW against man, animals, and crops has a long and detailed history. There are a number of examples in the historical record which suggest that the use of deliberate disease as a means of waging war has been in use since antiquity.[4] The literature on BW includes accounts relating to *ad hoc* attempts to deliberately spread disease through the contamination of water supplies and date back to 300 B.C.[5] This tactic of warfare, and the use of catapults to propel diseased human cadavers into besieged army fortifications, appear throughout the historical record until as late as 1763.[6]

During the latter half of the eighteenth century BW had become disease-oriented with both the British and the French attempting to introduce smallpox as a form of deliberate disease during conflicts against native American Indians.[7]

During the first World War, Germany was accused of spreading cholera in Italy and plague in St Petersburg,[8] and there is evidence that

German agents had conducted covert anti-animal BW sabotage operations infecting animals with anthrax and glanders causing disruption to American shipments of horses to the theatre of war in Europe.[9] In what can be regarded as an early example of anti-crop BW, grain shipments were also the subject of German sabotage operations during the First World War.[10]

The late nineteenth and early twentieth century saw the scope of BW expand considerably. The 'Golden Age' of bacteriology had emerged with scientists such as Pasteur and Koch demonstrating the causal relationship between pathogenic micro-organisms and disease,[11] and the potential of deliberate disease as a form of warfare against man, animals and plants had become the subject of systematic scientific study by the early part of the twentieth century.

Since scientifically-based BW became possible a number of important states have conducted detailed scientific investigations into spreading deliberate disease as a form of warfare. As well as evidence of German biological warfare activities during the First World War, French records show that systematic scientific studies into the potential of biological warfare began in France in the early 1920s.[12]

British and North-American programmes included investigations into both chemical and anti-crop BW with the American programme resulting in a well established and mature programme of anti-crop BW between the 1940s and the late 1960s.

During the late inter-war years offensive BW had progressed in Japan to the point where field-trials with anti-crop BW agents were conducted.

France conducted pre-war investigations into anti-crop BW and, in war-time Germany, investigations into anti-crop BW were pursued with some vigour.

Evidence regarding the proliferation of this form of warfare reminds us that it would be a mistake to assume that such weapons will never again emerge and that we can safely afford to ignore their development as a future possibility. The book concludes by considering ways in which this form of warfare might be most effectively controlled in the future.

Sources and approach

The data in this book has been compiled between 1994 and 1999 from a wide variety of sources. Anti-crop BW investigations before, during, and after the Second World War, were conducted in respective states under conditions of great secrecy and although many years has elapsed since these activities have ceased much of the information that would be needed to compile a full account of such programmes remains classified. A limited number of accounts relating to anti-crop BW developments have, however, appeared in the public domain and where it has been possible these sources have been consulted in full. In this connection, a notable contribution to the literature on the history of BW can be found in the original SIPRI (Stockholm International Peace Research Institute's) series, *The Problem of Chemical and Biological Warfare*, published between 1971 and 1975.[13] These sources, and additional secondary sources, including information relating to developments in BW subsequently published by SIPRI will appear throughout the pages that follow.

In compiling this book it was apparent from the outset that an account of such programmes would rely significantly on submitting requests for, and successfully obtaining in part or in full, previously classified primary source documentation from the archives of respective states. In the case of France and Japan, no primary source documentation relating to such programmes was made available to this author and secondary sources have therefore been relied upon in building up a picture of anti-crop BW developments in these countries. In regard to German BW activities, the book will draw on a previously classified intelligence reports and secondary source documentation.

In the case of the UK and the US, an insight into some aspects of collaboration on matters relating to anti-crop BW and the subsequent development of the US anti-crop BW programme has been assembled from both secondary source documentation, and from partially and/or fully de-classified official documentation, which has thus far remained outside of the public domain.

The enduring secrecy that surrounds military investigations into anti-crop BW has placed limitations on the breadth and depth of this investigation. The book draws on the best and in some cases the only source material available on the anti-crop BW activities of the respective states addressed in this study. However, in this connection, there is enough source material to present a coherent account of such activities.

While the topic is inter-disciplinary, cutting across strategic studies and the biological sciences, the approach adopted throughout this book is historical, demonstrating the political and scientific momentum behind post-Second World War BW research and development in general and anti-crop BW in particular. It must be further noted that it has not been the intention here – nor would it have been possible for the reasons stated above – to offer a definitive historical account of the anti-crop BW activities of respective states. A cut-off point for the collection of data for inclusion in this book was November 1999.

The book is organised into the following chapters:

Chapters

Chapter 2 discusses the events which lead up to United Nations Special Commission on Iraq's (UNSCOM) post-Gulf War revelations regarding Iraq's BW programme. Starting in 1991, this section covers a period of some seven years and includes an overview of the relevant United Nations Security Council Resolutions relating to Iraq's obligations to disclose in full the details of its BW. This chapter includes an account of what was learnt about Iraq's attempts to acquire a capability to wage anti-crop BW.

Chapter 3 offers both specialist readers, and readers with a non-scientific background, a general overview of the phytopathological classification of the pathogens that are the cause of the most serious diseases in crops – including those pathogens that have been known to have been developed as agents of choice in anti-crop BW programmes. In order to place the significance of disease in crops in a global context, this section also investigates world-wide losses to crops of economic importance.

Chapter 4 includes an overview of the development of legal constraints on BW. This chapter begins with an overview of the origins of the prohibition against BW and describes its current prohibition the 1972 Biological and Toxin Weapons Convention (BTWC). The BTWC entered into force in 1975 but lacks a formal mechanism to monitor the compliance of States Parties. This has reduced BTWC to a gentleman's agreement undermining its ability to resolve allegations of non-compliance and to address growing concerns about the proliferation of BW capabilities.[14] This section includes a discussion on the significance of plant pathogens in relation to the current initiative to negotiate a legally binding compliance and verification protocol and strengthen the Convention.

A major section of this book has been devoted to an investigation into the extensive US anti-crop BW programme which ran from 1942 until President Nixon announced the unilateral US renunciation of offensive BW research and development in 1969. The US BW programme represents the only example of a substantial and mature programme with a significant anti-crop BW component for which we have considerable publicly available documentation. Chapter 5, therefore, places anti-crop BW in the US in the context of the historical underpinnings of the wider US BW programme. The above offers an insight into the relationship between politics and science during the course of this period and argues that political involvement in developments regarding this form of warfare were in evidence at the highest levels in the US. In this chapter key domestic institutional developments, and the individuals who played a major role in such developments are identified. This chapter concerns the organisation of BW in the US from the period of Isolationism in the late inter-war years to the full mobilisation of science in support of the war effort.

Chapter 6 concerns the perceived threat from Axis BW activities from the perspective of the US and includes an investigation into offensive anti-crop BW activities in Germany and Japan. A review of French anti-crop BW activities has also been included in this section.

Chapter 7 discusses the organisation of US anti-crop BW in the post-war period and traces the changes in perceptions of threat during the course of this period. In the light of such perceptions, this section includes a series of official reviews regarding US chemical, biological and radiological warfare (CEBAR) policy. This section includes a detailed discussion of the CEBAR recommendations put forward by the US Research and Development Board, the Haskins Committee, and the Stevenson Committee.

From its somewhat humble origins in 1942/43 the US BW programme saw rapid expansion during the latter stages of the Second World War. North Atlantic collaboration on matters relating to chemical warfare had set a precedent for subsequent collaboration on BW, and research collaboration, information sharing, and reciprocal visits between the respective branches of the allied armed forces concerned with work on anti-crop BW formed an important component of these arrangements. From information hitherto not made available in the public domain, Chapter 8 discusses aspects of Anglo-American collaboration on anti-crop BW. In this chapter, correspondence exchanged between the UK and the US on matters relating to anti-crop BW offers an insight into the progress of war-time research and development on the identification of suitable agents in both countries. Then in the post-war period the book draws on documentation that has hitherto remained outside of the public domain to discuss collaboration between the UK's Crop Committee which was formed in 1948, under the auspices of the UK's Ministry of Supply, to oversee matters relating to warfare against crops; and its counterpart, the US Biological Branch of the Chemical Corps. This chapter covers aspects of a 15-year period of North-Atlantic scientific and technical information-sharing on anti-crop warfare through to 1958 when the US announced the temporary phasing out of US anti-crop BW research and development. Additionally this chapter includes investigations relating to the development, production, and stockpiling of anti-crop BW agents.

Chapter 9 gives an account of the development of anti-crop BW warfare munitions.

Chapter 10 investigates the US perception of Russia and China as potential anti-crop warfare targets. In this chapter there is a significant reliance on previously classified documentation. The potential economic, social and political effects of given percentage reductions in the calorific intake of the human populations in the above countries is discussed, and a major section in this chapter is devoted to a detailed appraisal of the operational requirements for an anti-crop warfare attack on two of the principal rice-producing areas in China.

Chapter 11, the penultimate chapter, considers whether the findings support the claim made in the introductory remarks to this book – that since it has become possible to conduct systematic scientific investigations into BW all known programmes have included investigations into offensive anti-crop BW. This chapter discusses speculation concerning anti-crop BW investigation in the former Soviet Union and goes on to discuss the potential for the proliferation of this form of warfare.

Although US anti-crop BW weapons were never used, in the final chapter it is argued that US research and development with anti-crop BW demonstrated the potential effectiveness of this devastating form of warfare. Current concerns over the proliferation of BW, and concerns over the possibility of enhancing the virulence of plant pathogens through genetic engineering in the future, emphasise the need for a careful analysis of anti-crop BW and its control. It is argued in this chapter that the strengthening of the prohibition against BW should form part of a wider web of deterrence which will help in ensuring that pathogens that pose a threat to crops are being used safely and for permitted purposes. In this connection, Chapter 12 concludes with an overview of the current prohibitions and controls on plant pathogens that are of relevance to effective and efficient implementation of the BTWC Protocol. Such initiatives include the strengthening of bio-safety and the safe storage, handling, use, and transfer of pathogens under the Convention on Biodiversity (Biosafety Protocol), the United Nations Environment Programme Guidelines on Biosafety, and the International Plant Protection Convention (IPPC). Also of relevance to the efficient and effective implementation of a strengthened BTWC are international ad hoc and non-government initiatives concerned with the surveillance and monitoring of crop disease outbreaks that present a danger to human health and the environment.

Appendix I provides an insight into the extent of US progress on anti-crop BW by the end of the Second World War. Although this information pertains only to US research on anti-crop BW, it has been included in order to allow the reader to develop a further appreciation of the nature of systematic scientific investigations into this form of warfare. Indeed it will be seen from the following that differences between offensive and defensive investigations in this phase of the research and development process are indistinguishable.

In the immediate post-war period, published works on anti-crop warfare related research began to appear in the scientific literature. However, it was not until 1947 that Rosebury[15] identified *Phytophthora infestans*, the cause of potato blight, along with two other fungal plant pathogens, *Piricularia oryzae* and *Helminthosporium oryzae*, respectively the cause of 'rice blast' and 'brown-spot' of rice, as agents mentioned in technical reports produced by Camp Detrick that might be *useful in biological warfare against plants* emphasis added. In connection with the study addressed in this appendix titled 'Studies on Factors Affecting the Infectivity of *Helminthosporium oryzae*',[16] it is interesting to note that E.C. Tullis, one of the co-authors of this 1947 paper, was also the senior author of the major 1958 study[17] published by Fort Detrick under the title, *The Importance of Rice and the Possible Impact of Anti-rice Warfare*. This case study, which was declassified in its entirety in April 1961, is discussed in some detail in Chapter 10 and provides a primary insight into thinking about how best to attack plant crops in China, and therefore also into the particular characteristics that such an attack might have.

2
Iraq and UNSCOM

In the years following the Gulf war, a wide-ranging process of investigation of Iraqi programmes for the development of weapons of mass destruction (WMD) was instigated under the auspices of the United Nations. UNSCOM was established in April 1991 by Security Council Resolution 687[1] and comprised some 150 staff (including Commissioners and staff) distributed between the United Nations Headquarters in New York, a field office in Bahrain, and the Monitoring and Verification Centre in Iraq. Assisted by an analytical support group known as the Information Assessment Unit, the Commission was organised into four distinct components: The Nuclear Group, the Chemical–Biological Group, the Ballistic Missile Group and the Long-Term Compliance Monitoring Group.

Security Council Resolution 687, the fourteenth UN Resolution adopted on Iraq since its invasion of Kuwait, gave the Commission a mandate to destroy Iraq's weapons of mass destruction affirming that in the interests of international peace and security in the area, Iraq should: [2]

> unconditionally accept the destruction, removal, or rendering harmless, under international supervision, of: (a) [a]ll chemical and biological weapons and all stocks of agents and all related subsystems and components and all research, development, support and manufacturing facilities; [and] (b) [a]ll ballistic missiles with a range greater than 150 kilometres and related major parts, and repair and production facilities.

Security Council Resolution 687 invited Iraq to reaffirm its obligations under the 1968 Treaty on the Non-Proliferation of Nuclear Weapons

(NPT). It also allowed for the imposition of sanctions and the use of military force under the provisions of Chapter VII of the Charter of the United Nations. Affirming the United Nations commitment to some 13 previous Security Council resolutions,[3] which placed a wide-ranging series of demands upon Iraq, Resolution 687 also held that Iraq should:[4]

> submit to the Secretary-General, within fifteen days of the adoption of the present resolution, a declaration of the location, amounts and types of all items specified [above], and agree to urgent, on-site inspections ... of Iraq's biological, chemical and missile capabilities, based on Iraq's declarations and the designation of any additional locations by the Special Commission itself.

Resolution 687 also charged the Commission with providing assistance and cooperation to the Director-General of the International Atomic Energy Authority (IAEA) in the inspection and destruction of Iraq's nuclear weapons production and maintenance facilities.

Resolution 707

In relation to the execution of its mandate the Commission was guided by two subsequent United Nations Security Council Resolutions which resulted from Iraqi non-compliance and failure to act in 'strict conformity with its obligations under Resolution 687.'[5] In addition to the stipulation set out in Resolution 687 regarding declarations, Resolution 707 demanded that Iraq: [6]

> (i) provide full, final and complete disclosure, as required by resolution 687 (1991), of all aspects of its programmes to develop weapons of mass destruction and ballistic missiles with a range greater than 150 kilometres, and of all holdings of such weapons, their components and production facilities and location, as well as all other nuclear programmes, including any which it claims are for purposes not relating to nuclear-weapons-usable material, without further delay,
>
> (ii) allow the special commission, the IAEA and their Inspection Teams immediate, unconditional and unrestricted access to any and all areas, facilities, equipment, records and means of transportation which they wish to inspect,
>
> (iii) cease immediately any attempt to conceal, or any movement or destruction of any material or equipment relating to its nuclear, chemical or biological weapons or ballistic missile programmes, or

material or equipment relating to its other nuclear activities without notification to and prior consent of the Special Commission.

and

(iv) make available immediately to the Special Commission, the IAEA and their Inspection teams any items to which they were previously denied access ...

Resolution 707 also established the right of the United Nations to perform other activities related to the inspection of research, development, production and maintenance facilities that were thought to be related components of Iraq's programme of weapons of mass destruction. Resolution 707 demanded that Iraq:[7]

allow the Special Commission, the IAEA and their Inspection Teams to conduct both fixed-wing and helicopter flights throughout Iraq for all relevant purposes including inspection, surveillance, aerial surveys, transportation and logistics without interference of any kind and upon such terms and conditions as may be determined by the Special Commission, and to make full use of their own aircraft and such airfields in Iraq as they may determine are most appropriate for the work of the Commission.

Resolution 715

A further Security Council Resolution, Resolution 715, demanded that Iraq submit to plans[8] for 'on-going monitoring and verification'[9] of Iraq's dual-use facilities that had been formulated and submitted for approval to the Security Council by the Secretary General and the Director General of the IAEA. Under Resolution 715 Iraq also had, *inter alia* to submit declarations regarding imports of items of dual-use nature and meet demands to:[10]

meet unconditionally all its obligations under the plans approved by the present resolution and cooperate fully with the special Commission and the Director General of the International Atomic Energy Agency in carrying out the plans.

The Commission conducted its first inspection of Iraq's facilities for the production of weapons of mass destruction in May 1991 relying mainly upon on-site inspections and overflights of suspected proscribed nuclear, missile, and CBW facilities. Prior to 1994 a considerable amount of attention had been paid to the inspection and destruction of nuclear weapons related facilities and the important power projection part of Iraq's long-range weapons capability. The destruction of Iraq's ballistic missile arsenal during the period to 1994 was therefore afforded a high priority.

Between the date of the first inspection and 1995 the Commission's inspection teams had also supervised the destruction of thousands of chemical munitions and chemical weapons-related components and production and maintenance facilities.

By 1995 inspections of suspected biological warfare facilities had risen above the number of inspections in the area of Iraq's ballistic missile capability. While the pattern of obstructing the inspection teams and Iraqi non-compliance persisted for a period of some eight years, a major breakthrough regarding investigations to reveal the scope and extent of Iraq's biological warfare programme resulted from the implementation of Resolution 715 and its final acceptance by Iraq in November of 1993, thus opening the way for the full implementation of the plans for ongoing monitoring and verification of Iraq's compliance with Security Council Resolution 687.

More thorough Iraqi declarations detailing imports of dual-use items, equipment and materials, revealed small quantities of complex growth media and dual-use equipment. Painstaking analysis of the quantities and recipients of such imports between late 1994 and early 1995 led UNSCOM inspectors to conclude that Iraq had successfully imported in the region of 38 tonnes of complex (biological warfare agent) growth media from a single source in two large shipments. This finding subsequently resulted in Iraqi acknowledgement of an offensive biological warfare capability.

Aspects of Iraq's programme of weapons of mass destruction[11] finally reached the public domain and were published in a report by UNSCOM in October 1995.[12] Where necessary, in the paragraphs that follow, the information in UNSCOM's report will be supplemented by information regarding Iraq's programme of weapons of mass destruction that has subsequently reached the public domain.

UNSCOM

Although the 1995 report covered elements of the nuclear weapons and missile programmes, it was noteworthy primarily as a remarkable account of the range and extent of the Iraqi chemical and biological warfare (CBW) programme from the early 1980s through to January 1991.

Nuclear and ballistic missiles programmes

As discussed above, primary responsibility for the destruction of Iraq's nuclear weapons capability, had been mandated to the IAEA. The following discussion, therefore, will comment on this aspect of Iraq's proscribed weapons programme only in the context of UNSCOM's findings with regard to ballistic missiles. While previous investigations had failed to reveal the extent of the proscribed missile programme and its relationship with other aspects of Iraq's programme of weapons of mass destruction, Iraq acknowledged information relating to the development and testing of missile systems that had previously remained undisclosed for a period of some four years. According to UNSCOM:[13]

> Iraq acknowledged for the first time work on advanced rocket engines, including those with increased thrust or using UDMH fuel. Iraq also admitted to the production of proscribed rocket engines made of indigenously produced or imported parts and through cannibalization of the imported Soviet-made Scud engines. Iraq further admitted that the number and the purpose of static and flight tests of proscribed missiles had previously been misrepresented ... [according to the Commission such tests included] both static and flight testing of Scud variant missile systems; several new designs of longer-range missile systems; development and testing of new liquid-propellant engine designs; development and successful testing of a warhead separation system; an indigenous design of a 600 mm diameter supergun system. ... Some of the previously undisclosed designs included missiles that could reach targets at ranges of up to 3,000 kilometres. The Commission also obtained information of a special missile under design for delivery of a nuclear explosive device.

Chemical warfare (CW) agents

Early investigations by the Commission's inspectors began in the summer of 1991. By the end of 1991 UNSCOM had identified a major chemical weapons manufacturing facility at Al Muthanna. Earlier

UNSCOM reports had indicated an Iraqi CW capability based on nerve agents such as sarin and tabun, and blister agents such as mustard gas. In relation to chemical warfare agents, Pearson has subsequently revealed that during the course of their initial investigations the inspectors had found evidence of the following CW agents:[14]

> Mustard, GB/GF, o-sec.butyl sarin (GS), n-butyl sarin, ethyl sarin (GE), spoiled tabun (GA), decomposed VX, and CS.

Declarations to the Commission also showed that Iraq had developed the potential to produce more of the advanced nerve agent VX than had previously been thought to be the case. In 1995 UNSCOM reported as follows:[15]

> Based on the new findings, it is now clear that the VX programme began at least as early as May 1985 and continued without interruption until December 1990. The Commission has concluded that VX was produced on an industrial scale. Precursor and agent storage and stabilization problems were solved. Furthermore, one of Iraq's documents on this subject, dated 1989, proposes 'the creation of strategic storage of the substance (VX-hydrochloride, one step from conversion into VX) so it can be used at any time if needed.
>
> Significant in this context is Iraq's admission; in September 1995, of the production in 1990 of 65 tonnes of choline,[16] a chemical used exclusively for the production of VX. This amount would be sufficient for the production of approximately 90 tonnes of VX. Furthermore, Iraq had, *inter alia*, over 200 tonnes each of the precursors phosphorous pentasulphide and di-isopropylamine. These quantities would be sufficient to produce more than 400 tonnes of VX.

Testing and weaponisation of CW agents

In addition to this wide-ranging CBW development programme, Iraq had conducted field trials to test its means of dissemination and subsequently moved on to weaponise a number of systems. According to the 1995 UNSCOM report, Iraq had, by January 1991, weaponised chemical and biological warfare systems in the form of bombs and missile warheads. With regard to chemical warfare agents the report reveals that:[17]

> Iraq [had] also admitted to the development of prototypes of binary sarin-filled artillery shells, 122mm rockets and aerial bombs.

> However, the new documentation show[ed] production in quantities well beyond prototype levels. Iraq [had] also admitted three flight tests of long-range missiles with chemical warheads, including one, in April 1990, with sarin.

The technologies associated with Iraq's CW delivery systems are believed not to be indigenous to Iraq. The report reveals that the above developments resulted from considerable assistance in the form of technical support and personnel from abroad.[18]

BW agents

Where previous declarations to UNSCOM revealed that Iraq had conducted BW research and development since the mid-1980s, until 1995, the extent of the programme had remained unconfirmed. More recent estimates report that by 1990 Iraq had produced large quantities of BW agent. According to UNSCOM:[19]

> In March 1988, a new site for biological weapons production was selected at a location now known as Al Hakam. ... In 1988, a search for production equipment for the biological weapons programme was conducted in Iraq. Two 1,850-litre and seven 1,480-litre fermenters from the Veterinary Research Laboratories were transferred to Al Hakam in November 1988. ... At Al Hakam, production of botulinum toxin for weapons purposes began in April 1989 and anthrax in May 1989. Initially much of the fermentation capacity for anthrax was used for the production of anthrax simulant for weapons field trials. Production of anthrax itself, it is claimed, began in earnest in 1990. In total, about 6,000 litres of concentrated botulinum toxin and 8,425 litres of anthrax were produced at Al Hakam during 1990.

While the significance of the figures relating to the above stockpile remain open to a certain amount of speculation, both botulinum toxin and anthrax are relatively well-known as potential anti-personnel BW agents. The UNSCOM report went on to indicate other areas of Iraq's BW research and development programme.

By 1988, this had been expanded to include an anti-personnel agent that causes a condition known as gas gangrene, as well as another, ricin. According to UNSCOM:[20]

> In April 1988, in addition to anthrax and botulinum toxin, a new agent, *Clostridium perfingens* ... was added to the bacterial research work at Al Salman. (*Clostridium perfringens* produces a condition known as gas gangrene, so named because of the production of gaseous rotting of flesh, common in war casualties requiring amputation of limbs).

Possibly unique in the development of BW agents, the Iraqi programme also included work on aflatoxin, a fungal toxin produced by species of *Aspergillus*, which is an animal and human carcinogen.[21] According to UNSCOM:[22]

> In May 1988, studies were said to be initiated at Al Salman on aflatoxin. (Aflatoxin is a toxin commonly associate with fungal-contaminated food grains and is known for its induction of liver cancers. It is generally considered to be non-lethal in humans but of serious medical concern because of its carcinogenic activity.). ... Research was conducted into the toxic effects of aflatoxins as biological warfare agents and their effect when combined with other chemicals. ... A total of about 1,850 litres of toxin in solution was declared as having been produced.

Testing and weaponisation[23] of BW agents

Field trials, and the testing of offensive BW agents and the means of dissemination for Iraq's anti-personnel BW weapons of mass destruction, were said to have taken place between March 1988 and 1990. After initial testing with aerial bombs using both anthrax simulant, *Bacillus subtilis*, and botulinum toxin, and a break in the testing programme that was claimed to have lasted some 18 months, testing of 122 mm rockets filled with anthrax simulant, botulinum toxin and aflatoxin were conducted in November 1989. According to UNSCOM:

> These tests were ... considered a success. Live firing of filled 122 mm rockets with the same agents were carried out in May 1990. Trials of R400 aerial bombs with *Bacillus subtilis* were first conduced in mid-August 1990. Final R400 trials using *subtilis*, botulinum toxin and aflatoxin followed in late August 1990. ... In December 1990, a programme was initiated to develop an additional delivery means, a biological weapons spray tank based on a modified aircraft drop tank. The concept was that tanks would be fitted either to a piloted

> fighter or to a remotely piloted aircraft to spray up to 2,000 litres of anthrax over a target. The field trials for both the spray tank and remotely piloted vehicle were conducted in January 1991. ... Three additional spray tanks were modified and stored, ready for use.

Field trials and testing of agent-filled BW weapons moved to weaponisation in the latter half of 1990. According to UNSCOM:[24]

> Weaponisation of biological warfare agents began on a large scale in December 1990 at Muthnanna. As declared, the R400 bombs were selected as the appropriate munition for aerial delivery and 100 were filled with botulinum toxin, 50 with anthrax and 16 with aflatoxin. In addition, 25 Al Hussein warheads, which had been produced in a special production run since August 1990, were filled with botulinum toxin (13), anthrax (10) and aflatoxin (2)
>
> In summary, Iraq has declared the production of at least 19,000 litres of concentrated botulinum toxin (nearly 10,000 litres were filled into munitions), 8,500 litres of concentrated anthrax (some 6,500 litres were filled into munitions) and 2,200 litres of concentrated aflatoxin (1,580 litres were filled into munitions)....These weapons were then deployed in early January 1991 at four location, where they remained throughout the war

Deployment

In an emergency deployment programme, immediately prior to the 1991 conflict, over 160 bombs and 25 Al Hussein missiles were deployed with chemical and biological warheads to four locations in Iraq. The biological components comprising botulinum, anthrax and aflatoxin warheads as described above. Furthermore, launch authority for use of these systems was pre-delegated to regional commanders. According to UNSCOM: [25]

> Iraq's intentions with regard to the operational use of its biological and chemical weapons have been subject to conflicting presentations by the Iraqi authorities in the period under review. On the one side, it was explained that the biological and chemical weapons were seen by Iraq as a useful means to counter a numerically superior force; on the other, they were presented as a means of last resort for retaliation in the case of nuclear attack on Baghdad.

> Certain documentation supports the contention that Iraq was actively planning and had actually deployed its chemical weapons in a pattern corresponding to strategic and offensive use through surprise attack against perceived enemies. The known pattern of deployment of long-range missiles (Al Hussein) supports this contention. Iraq stated, during visits of both the Chairman and the Deputy Chairman, that authority to launch biological and chemical warheads was pre-delegated in the event that Baghdad was hit by nuclear weapons during the Gulf War. This pre-delegation does not exclude the alternative use of such a capability and therefore does not constitute proof of only intentions concerning second use.

While the deployment of BW weapons during the Gulf War is a cause of concern one author has raised a question-mark over their effectiveness. According to Zilinskas,[26] the BW arsenal assembled by Iraq would have been ineffective for the following three reasons:

> (1) it was small; (2) payload dispersal mechanisms were inefficient; and (3) coalition forces dominated the theatre of war (ie, they had overwhelming air superiority and had crippled Iraq's command and control capability).

What the UNSCOM investigations demonstrate, however, is a substantial and systematic CBW research and development programme stretching over a number of years and leading to weaponisation at the outset of the war. The 1995 UNSCOM report also describes a specific initiative which suggested that Iraq had considered anti-crop BW as part of this overall process.

Plant disease epidemics as BW – the case of Iraq

In simple terms, plant pathogens that are deleterious to crop yields, are caused mainly by bacteria, viruses and parasitic fungi. While disease in plants is a matter to which we will return in the pages that follow (see Chapter 3, page 25), of the three groups, the Iraqi programme appears to have been concerned with fungi which cause a range of diseases known as rusts, blasts and smuts, specifically those affecting cereal crops such as rice, wheat, oats and corn. Of specific interest in the Iraqi programme was a disease known as cover smut or stinking smut or bunt of wheat, caused by fungi of the genus *Tilletia*, which causes crop losses in wheat in many parts of the world. The revelations regarding

Iraq's interest in what UNSCOM referred to as 'wheat cover smut' placed the development of such weaponry in the context of Iraq attempting to acquire a capability to wage so-called 'economic' warfare.[27] The potential impact of economic warfare with BW agents was described in a United States Air Force report[28] thus:

> Using BW to attack ... crops, or ecosystems offers an adversary the means to wage a potentially subtle yet devastating form of warfare, one which would impact the political, social, and economic sectors of a society and potentially of national survival itself.

While details of this element of Iraq's offensive BW programme remain rather limited, and according to Iraq, research and development stopped well short of weaponisation on any scale, it is understood that Iraqi BW workers had conducted limited testing on a small area field plot. According to UNSCOM:[29]

> After small production at Al Salman, larger-scale production was carried out near Mosul in 1987 and 1988 and considerable quantities of contaminated grain were harvested. The idea was said not to have been further developed; however, it was only sometime in 1990 that the contaminated grain was destroyed by burning at the Fudaliyah site.

While the intended use of this form of warfare remains open to a certain amount of speculation, it is probable, given the war with Iran, that Iraq's interest in cover smut was as a crop disease for use against Iranian wheat crops. According to the External Relations Unit at UNSCOM headquarters in New York:[30]

> Iraq [had] declared that it intended to use the smut as an economic weapon – their plan was to cause food shortages as part of a potential attrition war with Iran.

The following source gives an indication of the significance of the agricultural production of cereal crop in Iran:[31]

> the fundamental basis of Iranian agriculture is cereal production aimed at supplying local needs with wheat being by far the most important crop, though barley is [also] important. ... Rice is the only other significant cereal and the growth of this crop is mostly

> confined the Caspian Sea lowlands. These crops occupy between one half and three-quarters of the total cultivated area in any year.

It is conceivable, therefore, that Iranian agricultural production could have been vulnerable to this particular form of economic warfare.

Absence of further information relating to anti-crop BW developments in Iraq makes it difficult to provide more than an initial assessment of this aspects of Iraq's programme of weapons of mass destruction. Data relating to the actual size of Iraq's stockpile of *Tilletia* Spp are not available in the public domain, neither is any reliable data available with regard to the intended means of dissemination for wheat cover smut as a BW weapon. There is no available data relating to field-tests conducted on the small area field plot. It is not known what other plant pathogens, if any, were under consideration by Iraq's BW workers as potential anti crop BW agents. There is no data, over and above what has been included in the above text, relating to the intended targets. And absence of data relating to policy leaves us to speculate with regard to Iraq's intentions, and the perceived strategic and/or tactical utility of a capability to destroy crops as a form of BW.

UNSCOM inspections have successfully revealed, however, that Iraq had expressed an interest in acquiring the military capability to destroy crops as a form of BW and that considerable progress had been made with regard to research and development, leading to some form of weaponisation and testing. It will be noted that Iraq, therefore, appeared to have made some considerable progress in pursuit of an operational capability to conduct offensive anti-crop BW in accordance with the respective stages in the acquisitions process identified by the Office of Technology Assessment (see Table 1, Stage 5). Such developments formed part of a comprehensive chemical and biological warfare programme that focused mainly on developing anti personnel BW weapons and their means of dissemination. As discussed above, this resulted in alarming deployments of CW and BW weapons prior to the outbreak of war in the Gulf and the pre-delegation for use of such weapons to Iraq's field commanders.

We must now turn to develop a broader appreciation of the significance of disease in plants. The extent to which plant disease epidemics may be deliberately initiated as a form of BW will be dealt with in more detail in the chapters that follow.

3
The Study of Disease in Plants

Early plant disease epidemics

Historically, crop disease epidemics have, on occasions, caused massive economic losses and have even led to famines. The literature of plant pathology contains many references to natural outbreaks of plant disease epidemics. Disease in plants has also been shown to have resulted from human activity. The attendant socio-economic impact of such events has frequently been far reaching and is also well documented.

Biblical references to 'blasting' and 'mildew' provide an interesting historical indication of the destructive nature of plant disease epidemics and insect pests. Ancient Hebrews were thought to have regarded such calamities as God's punishment for man's wrongdoing. The following citation from I, *Kings*, is taken from Stakman and Harrar:[1]

> forgive the sin of thy servants and of thy people ... If there be in the land famine, if there be pestilence, blasting, mildew, locust, or if there be caterpillar; if their enemy besiege them in the land of their cities; whatsoever plague, whatsoever sickness there be; ... Then hear thou in heaven thy dwelling place, and forgive, and do, and give to every man according to his ways, whose heart thou knowest.

Disease of plants, legumes and trees was first studied and recorded at a somewhat observational and speculative level by the Greek philosopher Theophrastus (370–286 BC). The Romans attributed the destruction of grain crops (wheat and barley) to the rust god Robigus and according to Stakman and Harrar the ceremony of Robigalia was celebrated in the spring of each year from as early as 700 BC until, 'well into the Christian era'.[2] Little progress with regard to our understanding of the nature and causes of plant disease epidemics was achieved until well into the 17th century when according to Agrios, 'the invention of the compound microscope ... opened a new era in the life sciences'.[3]

In facilitating the study of disease in plants several criteria for the classification of such diseases subsequently emerged in the literature on plant pathology.

Definitions and classification of plant disease

First, however, Stakman and Harrar[4] offer a definition for what is meant by disease in plants:

> [a] plant disease is a physiological disorder or structural abnormality that is deleterious to the plant or to any of its parts or products, or that reduces their economic value. In this definition, 'physiological disorder' connotes any harmful deviation from normal physiology, regardless of cause.[5]

Causal groupings

The classification of disease in plants can be organised into a number of groupings. Disease in plants can be classified in accordance with the symptoms caused by the disease; or conversely, in accordance with the organ of the plant that is affected by the disease. Due to the diversity and large numbers of pathogens that affect plants, estimated to be in the region of tens of thousands, another useful criterion for classification identifies disease in plants in accordance with the types of plants affected.

An alternative classification of disease in plants, however, organises a diverse number of factors into respective groups, by identifying the cause of the disease in plants, as either: (1) pathogenic disease-producing organisms known as 'pathogens',[6] or by (2) physical environmental factors. In adopting this criterion for classification of disease in plants, it is therefore possible to make a distinction between infectious and so-called biotic or animate causes of disease in plants; and, non-infectious or so-called abiotic or inanimate causes of disease. The pathogenic group of factors, can therefore be categorised in accordance with the following classification adopted by Agrios,[7] as:

1. Diseases caused by fungi.
2. Diseases caused by bacteria.
3. Diseases caused by parasitic higher plants.
4. Diseases caused by viruses and viroids.
5. Diseases caused by nematodes. [and]
6. Diseases caused by protozoa.

With regard to the environmental or the non-pathogenic group of factors, we can again adopt the classification applied by Agrios[8] as:

1. Too low or too high temperature.
2. Lack or excess of soil moisture.
3. Lack or excess of light.
4. Lack of oxygen.
5. Air pollution.
6. Nutrient deficiencies.
7. Mineral toxicities.
8. Soil acidity or alkalinity (pH).
9. Toxicity of pesticides. [and]
10. Improper cultural practices.

While noting the significance of environmental factors in the classification of disease in plants, the discussion here will focus on the pathogenic group of disease producing organisms described above. Non-infectious, abiotic, and inanimate factors will be discussed, however, in relation to their role in plant disease outbreaks covered by the pathogenic group.

Two of the sub-categories included in the pathogenic group of phytopathogenic factors will subsequently also be excluded from the discussion that follows due to their relative insignificance when compared with pathogens that are considered responsible for disease in plants of important economic significance. The discussion will therefore be limited to: diseases caused by fungi; diseases caused by bacteria; diseases caused by viruses; diseases caused by nematodes; and diseases caused by insects. Each section will also include a brief overview of the symptoms caused by the disease, and the way in which its transmission most commonly occurs in the plant kingdom. Each respective section will conclude with an overview of disease outbreaks associated with the different pathogens.

Although the intention here is not to offer a comprehensive or definitive phytopathological interpretation of the role of pathogens in the plant kingdom, the discussion is intended to allow the reader to develop an appreciation of the nature and causes of disease in plants, and the way in which plant disease can sometimes spread resulting in significant losses to plant crops of international significance.

Some of the main characteristics of pathogens are therefore summarised in accordance with key reference works from the literature on phytopathology as follows:

Disease in plants

Pathogens may cause disease in plants in the following ways:[9]

> by (1) weakening the host by continually absorbing food from the host cells [of the plant] for their own use; (2) killing or disturbing the metabolism of host cells through toxins, enzymes, or growth-regulating substances they secrete; (3)blocking the transportation of food, mineral nutrients, and water through the conductive tissues [of the plant]; and (4) consuming the contents of the host cells [of the plant] upon contact.

Symptoms

Some common symptoms of disease in plants are categorised by Stakman and Harrar[10] as follows, as: (1) necrosis, (2) hypoplasia, and (3) hypertrophy:

> Necrosis is the rotting or decay of tissues; hypoplasia is underdevelopment; and hypertrophy is abnormal overgrowth.[11]

Fungi

In accordance with the classification adopted from Agrios the first infectious group are represented by fungi. According to Parry,[12] the general characteristics of fungi as plant pathogens are as follows:

> Fungi are the most important plant pathogens. Plant pathogenic fungi are microscopic organisms which, unlike plants, lack chlorophyll and conductive tissue. Of the 100,000 or more species of fungi ... more than 8000 species are known to cause disease in plants. ... A typical fungus consists of a vegetative body (mycelium) made up of individual branches (hyphae) which may or may not have cross-walls (septa). ... Fungi reproduce chiefly by means of

spores. Spores are specialised propagative or reproductive bodies, consisting of one or a few cells...

Fungi lack chlorophyll but require the carbohydrates, fats, proteins and nutrients that are dependent upon the process of photosynthesis in green plants in order to survive and reproduce, they are therefore dependent upon organisms, either living or dead, and are attributed the classification of either parasite[13] or saprophyte.[14]

Dissemination of fungi

In the spread of disease caused by fungal infection the dissemination of reproductive spores throughout crops is of key importance. While the initial discharge of spores from some fungi is active, so to speak, with some fungi forcibly discharging and disseminating reproductive spores, the overwhelming majority of fungi rely on the dissemination of reproductive spores by so-called 'agents of dissemination'. With regard to this form of dissemination, fungi are considered to be passive. According to Agrios:[15]

> The distance to which spores may be disseminated varies, with the agent of dissemination. Wind is probably the most important disseminating agent of spores of most fungi and may carry spores over great distances. For specific fungi, other agents such as water and insects may play a much more important role than wind in the dissemination of their spores.

Human activity is also responsible for the dissemination of fungal plant pathogens.

In a rare contribution from the literature on plant pathology one scientist offers an interesting insight into the potential of fungal plant pathogens as BW agents. According to Van Der Plank,[16] writing in 1963:

> We often call an epidemic explosive. In time of peace the adjective is neatly descriptive. In time of war it could be grimly real in the military sense. An enemy has few explosives to surpass a pathogen that increases at the rate of 40% per day, compounded at every moment, and continues to increase for several months.
>
> Spores are as light as poison gas or smoke. There are 150 billion uredospores of the wheat stem rust fungus to the pound: 3×10^{14} to the ton. Many types of spore disperse as easily as smoke. Many are

tough and durable. They have only to be dispersed in the proper places at the proper times. Nature sees to the explosion.

As Agrios[17] has written, the most common so-called necrotic symptoms of fungal disease of plants can be categorised as in Table 3.1.

Additionally, the most common symptoms of stunting (hypertrophy) and excessive growth (hyperplasia) in plants are included in Table 3.2.

Examples of crop losses due to fungi

One of the most calamitous recorded outbreaks of disease in plants caused by fungi is the Irish Famine caused by the failure of the potato crop in 1845–46. During the 1840s and 1850s and again during 1870s and 1880s, Europe experienced significant loss to its grape crops due to fungal plant disease epidemics. In the late 1800s (1870–1880), coffee rust was responsible for the destruction of the entire coffee crop in S.E. Asia, and during the course of the 20th century brown spot of rice

Table 3.1 Necrotic symptoms of fungal disease in plants

Leaf spots	Localised lesions on host leaves consisting of dead and collapsed cells
Blight	General and extremely rapid browning of leaves, branches, twigs, and floral organs resulting in their death.
Canker	A localised wound or necrotic lesion, often sunken beneath the surface of the stem of a woody plant.
Die-back	Extensive necrosis of twigs beginning at their tips and advancing toward their bases.
Root rot	Disintegration or decay of part or all of the root system of a plant.
Damping off	The rapid death and collapse of very young seedlings in the seed bed or field
Basal stem rot	Disintegration of the lower part of the stem
Soft rots and dry rots	Maceration and disintegration of fruits, roots, bulbs, tubers, and fleshy leaves
Anthracnose	A necrotic and sunken ulcer-like lesion on the stem, leaf, fruit, or flower of the host plant
Scab	Localised lesions on host fruit, leaves, tubers, etc. usually slightly raised or sunken and cracked, giving a scabby appearance
Decline	Plants growing poorly; leaves small, brittle, yellowish, or red; some defoliation and die-back present

Table 3.2 Hypertrophy and hyperplasia symptoms caused by fungal disease in plants

Clubroot	Enlarged roots appearing like spindles or clubs
Galls	Enlarged portions of the plants usually filled with fungus mycelium
Warts	Wart-like protuberances on tubers and stems
Witches'-brooms	Profuse, upward branching of twigs
Leaf curls	Distortion, thickening, and curling of leaves
Wilt	Usually a generalised secondary symptom in which leaves or shoot lose their turgidity and droop because of a disturbance in the vascular system of the root or of the stem
Rust	Many, small lesions on leaves or stems, usually of a rusty colour
Mildew	Chlorotic or necrotic areas on leaves, stems, and fruit, usually covered with mycelium and the fructifications of the fungus

was the cause of the devastating Bengal famine of 1943. Fungal plant pathogens remain the cause of frequent world-wide epidemics in plants in both advanced industrialised and developing countries resulting in huge annual losses. Currently, fungal diseases cause losses running into billions of dollars each year throughout the world, especially on cereals, vegetables and fruit.

Bacteria

The second infectious group are represented by bacteria. According to Agrios:[18]

> Bacteria are simple micro-organisms. ... About 1600 bacteria are known. About 80 species of bacteria ... have been found to cause diseases in plants. ... Bacteria may be rod shaped, spherical, ellipsoidal, spiral, comma shaped, or filamentous (threadlike). ... Some [bacteria] can transform themselves into spores, and certain filamentous forms can produce spores, called conidia, at the end of the filament. ... Bacteria multiply with an astonishing rapidity, and their significance as pathogens stems primarily from the fact that they can produce tremendous numbers of cells in a short period of time. Bacterial diseases of plant occur in every place that is reasonably moist or warm, they affect almost all kinds of plants, and under favourable environmental condition, they may be extremely destructive.

Symptoms associated with bacteria

With regard to the symptoms of disease in plants caused by bacteria, according to Agrios:[19]

> Plant-pathogenic bacteria cause the development of almost as many kinds of symptoms on the plants they infect as do fungi. They cause leaf spot and blights, soft rots of fruit, root, and storage organs, wilts, overgrowths, scabs, cankers, and so on. Any given type of symptom can be caused by bacterial pathogens in several genera, and each genus contains some pathogens capable of causing different types of diseases.

Dissemination of bacteria

There are four principle means for the dissemination of bacteria from one plant to another: water, insects, other animals and humans. Other factors also contribute to the distribution of bacteria: contaminated seed, infected transplants, and a variety of human activity including the action of tractors and ploughs, and the action of other implements used in the process of production and harvesting. Dissemination in water can be caused by irrigation and flooding, rain splash and water running off contaminated plants and soil, and by rain that is blown by wind. Water can be responsible for spreading bacteria from plant to plant, from one part of the plant to anther part of the plant, from plant to soil and or vice-versa, and from soil to underground parts of the plant. Some bacterial plant pathogens inhabit the plant or parts of the plant and do not thrive in soil while others inhabit soil and do not thrive in plants. In the paragraph that follows, Agrios[20] outlines the role of insects, other animals, and humans, in the dissemination of bacterial plant pathogens:

> Insects not only carry bacteria to plants, they also inoculate the plants with the bacteria by introducing them into the particular sites in plants where they can almost surely develop. In some cases bacterial plant pathogens also persist in the insect and depend on them for their survival and spread. In other cases, insects are important but not essential in the dissemination of certain bacterial plant pathogens. Birds, rabbits, and other animals visiting or moving among plants may also carry bacteria on their bodies. Humans help spread bacteria locally by handling plants and by their cultural practices, and over long distances by transportation of infected plants or plant parts to new areas or by introduction of such plants from

other areas. In cases in which bacteria infect the seeds of their host plants, they can be carried in or on them for short or long distances by any of the agencies of seed dispersal.

Examples of crop losses due to bacteria

In the 1980s Citrus canker, which also affects trees in Asia, Africa, and Brazil, affected millions of trees in Florida. Severe losses caused by Citrus canker were also experienced in Florida around the turn of 20th century (1910–1920). Fire blight of pome fruits causes considerable losses annually in North America and Europe.

Viruses

The third infectious group are represented by viruses. According to Agrios:[21]

> A virus is a nucleoprotein that is too small to be seen with a light microscope, multiplies only in living cells, and has the ability to cause disease. All viruses are parasitic in cells and cause a multitude of diseases to all forms of living organisms, from single-celled microorganisms to large plants and animals. About one-fourth of all known viruses attack and cause diseases of plants....Viruses do not divide and do not produce any kind of specialised reproductive structures such as spores, but they multiply by inducing host cells to form more virus. Viruses cause disease not by consuming cells or killing them with toxins, but by utilising cellular substances, taking up space in cells, and by disrupting cellular components and processes, which in turn upset the metabolism of cells and lead to the development by the cell of abnormal substances and conditions injurious to the function and the life of the cell or the organism.

Symptoms associated with viruses

The symptoms associated with virus infection can be more subtle than those caused by fungi and bacteria. However, this is a general rule and viruses can be the cause of obvious symptoms of infection. According to Agrios:[22]

> The most common and sometimes the only kind of symptom produced by virus infection is reduced growth rate of the plant, resulting in various degrees of dwarfing or stunting of the entire plant. Almost all viral diseases seem to cause some degree of reduction on

> total yield, and the length of life of virus-infected plant is usually shortened. The most obvious symptoms of virus-infected plant are usually those appearing on the leaves, but some viruses may cause striking symptoms of the stem, fruit, and roots, with or without symptom development on the leaves. ... The most common types of plant symptoms produced by systemic virus infection are *mosaics* and *ring spots.*

Dissemination of viruses

Like fungal plant pathogens, viruses may be disseminated from plant to plant, passively, by wind and water, but the most common means of dissemination is that of vectors. This form of dissemination requires that the virus enters the plant via some form of wound. A virus disseminated in this way can be passed from one generation of one plant to a subsequent generation. According to Agrios,[23] virus transmission from plant to plant occurs, in the following ways, via:

> vegetable propagation; mechanically through sap; and by seed, pollen, insects, mites, nematodes ... and fungi.

While the relationship between viruses and vectors is somewhat complicated, dissemination by insect is the most common form of virus dissemination. Amongst this category of vectors, aphids are responsible for the transmission of some 170 viruses, whilst leafhoppers are responsible for the transmission of some 40 plant viruses.[24]

Examples of crop losses due to viral diseases

Plum pox or sharka has been responsible for severe epidemics on apricots, peaches and plums throughout Europe. In Africa, Swollen shoot of cacao has been responsible for continuous heavy losses, while Citrus quick decline (tristeza) has killed millions of citrus trees. Sugarcane mosaic and Sugar beet yellows have been, and continue to cause, heavy losses to sugarcane and corn worldwide.

Nematodes

Nematodes are multi-cellular invertebrate worms:[25]

> Sometimes called eelworms, nematodes are wormlike in appearance but quite distinct taxonomically from the true worms. ... Several hundred species [of nematodes], however, are known to feed on

> living plants, causing a variety of plant diseases. ... Plant-parasitic nematodes are small, 300–1000 μm, with some up to 4 mm long, by 15–35 µm wide. Their small diameter makes them invisible to the naked eye, but they can be easily observed under the microscope. ... All plant-parasitic nematodes have a hollow stylet or spear, which is used to puncture plant cells.

Symptoms associated with nematodes

With regard to the symptoms associated with nematode infection:[26]

> Nematode infection of plants result in the appearance of symptoms on roots as well as on the above ground parts of plants. Root symptoms may appear as root knots or root galls, root lesion, excessive root branching, injured root tips, and root rots when nematode infection are accompanied by plant-pathogenic saprophytic bacteria and fungi.
>
> These root systems are usually accompanied by non-characteristic symptoms in the above ground parts of plants, appearing primarily as reduced growth, symptoms of nutrient deficiencies such as yellowing of foliage, excessive wilting in hot or dry regions, reduced yields, and poor quality of products.
>
> Certain species of nematodes invade the above ground portions of plants rather that the roots, and on these they cause galls, necrotic lesions and rots, twisting or distortion of leaves and stems, and abnormal developments of the floral parts. Certain nematodes attack grains or grasses forming galls full of nematodes in place of seed.

Dissemination of nematodes

Nematodes spend most of their life in soil and many are found in soil close to the roots of vulnerable plants. The nematode moves through the soil under its own steam, so to speak, and it is estimated that in one season it will not travel a distance greater than one meter. Not unlike fungi and bacteria, however, nematodes can also be carried by wind, water and by human activity and in this way they can attack the above ground parts of plants.

Examples of crop losses due to nematode diseases

Root knot has been responsible for causing considerable and continuous losses to vegetables and other plants worldwide and the Sugarbeet syst nematode has caused severe losses to sugarbeet in Northern Europe and Western United States.

Insects

It is apparent from the discussion so far that the role of insects is of particular significance to disease in plants and its transmission. According to Agrios, '[i]n the case of viruses ... and fastidious bacteria ... insects are the most important disseminating agents'.[27] While insects fall under the animate category of causes of disease in plants in the literature on phytopathology, they do so as the direct cause of disease, and as agents of dissemination, of bacteria, fungi and viruses. The most important insect vectors are aphids and leafhoppers with piercing and sucking mouthparts. Insects with piercing and sucking mouthparts act as a conduit for the virus from insect to plant.[28] An important distinction, however, separates the direct cause of disease by insects, including those that act as 'agents of dissemination', from the literature on economic entomology which concerns the physical destruction of the plants through chewing by the insects. The following insects are responsible for the dissemination of phytopathogenic agents:

Piercing and sucking insects

1 Aphids
2. Leafhopper
3. Psylla
4. Treehopper
5. Planthopper
6. Whitefly
7. Mealybug and Scale insects
8. Plant bug

Infestation of plants caused by the following also results in the physical destruction of parts of the plant.

Chewing insects

9. Thrips
10. Beetle
11. Grasshopper

Insects can undermine the quality and reduce crop yields and increase the costs of harvesting and production of crops due to the expense that is associated with their control. Insects can cause significant damage to both cereal and vegetable crops. The role of insects in the direct and indirect destruction of crops of important international significance is a matter to which we will return.

Economic crops of international importance

In order to restrict the scope of the topic somewhat further, disease in plants will be discussed in relation to a selected group of economic crops of international importance. The following discussion, therefore, follows the approach adopted by Stakman and Harrar by categorising disease in plants in accordance with the types of plants affected. This concerns three groups of plants: cereals, potatoes and tropical fruits. However, due their international significance, sugar and coffee will be included in the discussion.

Cereal crops

It has been estimated that cereal crops contribute roughly 80 per cent of the worlds calorific requirements.[29] The crops of major international significance include wheat, rice and maize.

Wheat (Triticum vulgare)

A common form of disease in wheat is rust fungi. According to Stakman and Harrar,[30] rust fungi, 'cause greater annual losses in wheat production than all the other pathogenic organisms attacking wheat.' However, other pathogens are responsible for causing considerable losses with regard to the production of wheat crops. Most prominent among these pathogens are several species of smuts.

Some of the most important pathogens affecting wheat are listed below (Table 3.3).

Table 3.3 Important pathogens affecting wheat

Puccinia graminis tritici (fungi)	Stem Rust of Wheat
Puccinia glumarum (fungi)	Stripe rust
Puccinia rubigo-vera (fungi)	Leaf rust
Ustilago tritici (fungi)	Loose smut
Tilletia spp (*Iraq's choice of pathogen as biological warfare agent* (fungi))	Stinking smut (*referred to by UNSCOM as 'wheat cover smut'*)
Urocystis tritici (fungi)	*Flag smut*
Gibberella zeae (fungi)	Scab
Septoria nodorum (fungi)	Blotch
Ophiobolus graminis (fungi)	Take-all

Rice (*Oryza sativa*)

The production and consumption of rice occurs on a scale of worldwide significance. According to Chapman and Carter, '[r]ice is the most important food crop and the major dietary component for more than a third of the world's population. Rice production is concentrated in Asia where more than 90% of the world's supply is produced'.[31] Some of the most important pathogens affecting rice are listed below (Table 3.4).

Maize/Corn (*Zea mays*)

The production and consumption of maize is also of considerable significance. According to Chapman and Carter, '[c]orn is the only important cereal thought to have evolved in the New World, reportedly in Mexico ... [and] is the leading feed crop in the United States'.[32] Maize is susceptible to a wide variety of pathogens. The most significant of these pathogens are listed below (Table 3.5).

Potatoes (*Solanum tuberosum*)

Since its introduction from the New World the potato has become a major food crop in Europe. According to Chapman and Carter, 'potatoes rank second only to cereal crops in importance'.[33] The potato is host to numerous plant pathogens, including some destructive plant viruses. Some of the most significant fungal and bacterial pathogens are listed below (Table 3.6).

Tropical fruits

Banana (Musa domestica)

Banana is one of the most important food crops of the world. According to Stakman and Harrar:[34]

Table 3.4 Important pathogens affecting rice

Piricularia oryzae (fungi)	Rice blast
Helminthosporium oryzae (fungi)	Brown spot

Table 3.5 Important pathogens affecting maize

Puccinia sorghi (fungi)	Maize rust
Helminthosporium maydis (fungi)	Leaf blight
Phycomycete sclerospora (fungi)	Downy mildew
Virus spread by *Cicadulina mbila*	Streak

Table 3.6 Important pathogens affecting potatoes

Phytophthora infestans (fungi)	Late blight
Streptomyces scabies (fungi)	Common scab
Spongospora subterranea (fungi)	Powdery scab
Synchytrium endobioticum (fungi)	Wart
Pseudomnas solanacearum (bacteria)	Brown rot
Corynebacterium spedonicum (bacteria)	Ring rot
Alternaria solani (fungi)	Early blight
Verticillium albo-atrum (fungi)	Wilt
Erwinia atroseptica (bacteria)	Blackleg

It is grown almost universally throughout the tropics and subtropics, is highly prized as a food in the areas of production, and is in demand on non-producing countries. The production and export of bananas is one of the leading agricultural industries in Colombia, Ecuador, Honduras, Costa Rica, Guatemala, Panama, Jamaica, and parts of southern Europe. Any disease which adversely affects this crop may have disastrous effects on the economy of the banana-producing countries.

Banana is host to numerous plant pathogens but most are not recorded as having caused disease in bananas of epidemic proportions. Two plant pathogens of major international significance involved in the destruction of banana crops are as follows (Table 3.7).

Citrus fruits (*Citrus spp*)

Citrus fruits and their related products are numbered among the most valued international food commodities'.[35] Citrus fruits are host to numerous pathogenic fungi, bacteria, viruses and nematodes, but the most damaging pathogens to the production of citrus are (Table 3.8).

Table 3.7 Important pathogens affecting bananas

Cercospora musae (fungi)	Leaf spot
Fusarium oxysporum f. cubense (fungi)	Wilt

Table 3.8 Important pathogens affecting citrus

Psorosis disease caused by numerous viruses	Tristeza or quick decline
Phytophthora citrophthora (fungi) (and *Phytophthora parasitica*) (fungi)	Foot rot

Sugar cane (*Saccharum officinarum*)

Sugar cane has been a crop of major economic significance since the latter half of the eighteenth century. Sugar cane is significantly affected by the following viral plant pathogens: Sereh, Mosaic, Fiji, streak, dwarf, chlorotic streak and ratoon stunting. Some of the most destructive plant pathogens to effect sugar cane which include fungi, bacteria, and nematodes are summarised below (Table 3.9).

Coffee (*Coffea spp*)

As a principle commodity of over 50 countries coffee production represents one of the worlds major export crops. Some plant pathogens of major significance are as follows (Table 3.10).

While it would be incorrect to underplay the significance of non-infectious, abiotic, and environmental factors with respect to their role in disease in plants, it will be clear from the preceding discussion that a number of plant pathogens have been capable of causing disease outbreaks of considerable international significance, with fungal plant pathogens being responsible for some of the most calamitous losses to crop production, particularly in the case of cereal crop (wheat and rice), and potatoes.

Table 3.9 Important pathogens affecting sugar

Ustilago scitaminea (fungi)	Sugar cane smut
Colletotrichum falcatum (fungi)	Red rot
Xanthomanas vasculorum (bacteria)	Gummosis
Helminthosporium sacchari (fungi)	Eye spot
Sclerospora sacchari (fungi)	Downy mildew
Xanthomanas albilineans (bacteria)	Leaf scald
Marasmius sacchari and *Pythium arrhenomanes* (fungi)	Root rots
Gibberalla fujikoroi (fungi)	Stalk and seed cane rot

Table 3.10 Important pathogens affecting coffee

Hemileia vastatrix (fungi)	Rust
Omphalia flavida (Stilbella flavida (fungi))	Leaf spot
Corticium salmonicolor (fungi)	Pink disease
Cercospora (fungi)	Leaf spot
Cercospora coffeicola (fungi)	Brown eye spot
Pellicularia koleroga (fungi)	Thread blight
Gloeosporium cingulatum and Colletotrichum coffeanum (fungi)	Anthracnose or leaf drop

Crop disease in global perspective

Estimated losses[36] to world crop production suggest that disease in plants continues to be a problem of considerable significance. The following statistics provide a brief overview of the extent of the problem and some possible implications. The data presented below is intended to illustrate, in general terms, the high level of crop losses to disease in developing countries, when compared to developed countries. In the context of comparatively high levels of crop losses to disease in developing countries, this data is set against a background of data projections that suggest that the rising populations of developing countries will place and increasing burden on food production where levels of crop losses to disease are high. This data is intended to be illustrative and not definitive. On the basis of this limited data, this discussion will suggest that crop production in developing countries may become increasingly vulnerable if the international prohibition against anti-crop BW remains weak.

One particular estimate suggests that pre-and post-harvest losses to disease and insects amounted to a staggering 48 per cent of all potential food crops in 1982.[37] Taking each particular crop in isolation, the following crops loss estimates to cereal, potato, other root crops, sugar cane, vegetables, fruits, and coffee–cocoa–tea, are based upon the standard publication –*Estimated 1982 Crop Production and Preharvest Losses (in Millions of Tons) and Percent Lost to Diseases and Other Pests (Insects, Weeds) in Developed and Developing Counties*, cited in Agrios.[38]

Losses to world crop production

Out of an estimated total potential production of 2,588 million tons of world cereal production in 1982, 9.2 per cent of cereal was lost to diseases, and 13.9 per cent of cereal was lost to insects. Of 376 million tons of potential world production of potatoes, some 21.8 per cent of potatoes were lost to diseases, whilst 6.5 per cent of potatoes were lost to insects. Disease amounted to some 16.7 per cent of losses to potential world production of other root crops, with insects accounting for some 13.6 per cent of losses out of a total potential world production of some 976 million tons. Out of a total potential world production of sugarcane of 1,802 million tons, disease accounted for some 19.2 per cent of losses, whilst insects accounted for some 20.1 per cent of losses. Disease accounted for some 10.1 per cent of losses to vegetables according to 1982 estimates, and insects accounted for some 8.7 per cent of losses. Out of a potential world production of 394 million tons of fruits, 12.6 per cent of losses were accounted for by diseases, whilst 7.8 per cent of losses were accounted for by insects. Of 15 million tons of potential world production of coffee–cocoa–tea, some 17.7 per cent of losses were accounted for by diseases, and 12.1 per cent of losses were accounted for by insects.

Crop production in developed and developing countries

A number of rather interesting characteristics with regard to the above estimates arise from the distinction that can be drawn between crop losses caused by disease in countries of the developed world (Europe, North America, Australia, New Zealand, Japan, Israel and South Africa), and the loss to crops caused by disease in developing countries. In 1982, out of a population of some 1,186 million, only 11.6 per cent of the population of developed countries were engaged in agricultural production (Table 3.11). While in developing countries, with an estimated population of some 3,405 millions, the percentage of the population involved in agricultural production was significantly higher at some 57.6 per cent (Table 3.12).

As a percentage of the total yield developing countries loose more of the above crops to disease than developed countries. The estimated percentage of greater losses for the six food types in developing countries are illustrated in the following table (Table 3.13). This is

Table 3.11 Crop losses to disease in developed countries[39]

Developed countries
Population (million) = 1,186
% in agriculture = 11.6
Arable land (millions hectares) = 672,335

Crops	*Actual production*	*Crop losses (million tons) Millions of tons lost to disease*
Cereals	874	51
Potatoes	204	40
Other root crops	205	25
Sugar cane	77	14
Vegetables	145	16
Fruits	138	17
Coffee–cocoa–tea	0	–

Table 3.12 Crop losses to disease in developing countries[40]

Developing countries
Population (millions) = 3,405
% in agriculture = 57.6
Arable land (Million hectares) = 796,264

Crops	*Actual production*	*Crop losses (million tons)*
Cereals	822	187
Potatoes	51	21.6
Other root crops	350	138
Sugar cane	734	332
Vegetables	223	36
Fruits	164	32
Coffee–cacoa–tea	8	2.5

Table 3.13 Comparison of ratio lost in developing countries compared to developed

Crop/product	*% Greater loss of developing countries over developed countries*
Cereal	91.3
Potato	20.4
Other root crops	28.3
Sugar	49.7
Vegetables	46.3
Fruits	58.5

particularly striking given that developing countries have significantly higher populations than that of the developed countries, yet produce relatively less in terms of actual production of food crops, and suffer more from losses to disease.

Although a detailed analysis of crop disease estimates is outside of the purview of this study, more recent data relating to crop losses in developed and developing countries supports the estimates for losses to crops set out by Agrios above. Based upon estimates which compare developed and developing between 1965 and 1990, Oerke *et al.*[41] arrive at broadly similar conclusions to Agrios. Thus they argue that:

> Since 1965, total loss for most crops have remained constant or even decreased in the developed countries, especially in West

> Europe. In most developing countries, despite an increase of actual yields, loss rates have increased, considerably in some regions, because attainable yields were raised due to the use of high yielding varieties, improved irrigation and higher fertilizer amounts, and crop protection has not been improved to an appropriate extent ...
>
> West Europe where farmers harvest about 77% of the potential, contributes only 8% to the actual global production of ... crops, whereas Africa, Latin America, Asia, East Europe and the U.S.S.R. where losses reach 45%, account for 81% of the potential, and 77% of the actual production.

Oerke *et al.*[42] go on to offer an explanation for regional variations in crop losses thus:

> There are several reasons for the bias in loss rates among the regions. In the tropics and subtropics, the incidence and diversity of harmful organisms is higher and the climate favours the epidemics of pathogens and animal pests, in semi-arid regions weed competition is higher than under humid conditions. As many tropical crops are highly susceptible to pests, and, due to the developmental costs, the use of resistant varieties is often confined to industrialised countries ...

Such estimates paint an alarming picture when viewed in the context of projected population increases, which suggest that the populations of developing countries will rise from some 4,204 million in 1990, to some 5,473 million in the year 2000.[43] Estimates for population increases in the developed countries where there is less of a reliance on agricultural production, and less damage to crops caused by disease, are much more modest, with the population at 1990 set to rise from an estimated 1,266 million, to some 1,347 million in the year 2000. If the trends suggested by the above statistics continue, a significant increase in the populations of developing countries will place a greater burden on crop production where losses to disease are already relatively high. Again the data relating to projected increases in global population presented by Agrios are supported by further data presented by others. Based upon estimates made by the World Bank Oerke *et al.*[44] argue that:

> [Between] 1985 and 2025 the world's population will increase by 3,571 million people, from 4,844 to 8,415 million. Developing countries will account for 96% of this increase, which will add

> another 3,414 million to the 1985 population of 3,666 million in the developing countries ... by the year 2050 the population of the developed countries will be a mere 1,319 million, or 13% of the total population of the world.

Given the extent of estimated crop losses to disease, and the increasing burden of population increase in regions dominated by developing countries, such countries would appear to be particularly vulnerable to non-accidental or non-natural outbreaks of disease in crops of important economic significance.

In the absence of further information relating to Iraq's anti-crop BW programme the literature on plant pathology discussed above shows reasonably conclusively that disease in plants has been responsible for some calamitous crop failures with far reaching social and economic implications and that, in spite of advanced methods of control, damage caused by plant pathogens remains a considerable problem, particularly with regard to the production of crops of important international significance such as cereal, potato, tropical fruits and other cash crops. Moreover, recent estimates regarding projected increases in population in developing countries, if correct, would seem to compound the problem of losses to crop production from disease somewhat further.

The work to acquire a capability to wage anti-crop BW in Iraq took place in parallel to negotiations that seek to strengthen the international community's resolve to outlaw the hostile use of biomedical technologies. The process to negotiate a legally binding compliance regime for the Biological Weapons Convention in the form of a Protocol is of particular significance to the prevention and non-proliferation of anti-crop BW and it is to this matter that we will now turn.

4
Anti-Crop BW and the BTWC

Development of the legal prohibition against BW

According to Moon,[1] the use of biological weapons in war has been the subject of restrictions since the principle prohibiting their use was established in customary law[2] dating back to the classical Greek and Roman period. The restrictions on the use of BW in customary law was subsequently defined in the eighteenth century, and codified in the century that followed. According to Dando:[3]

> The International Declaration concerning the Laws and Customs of War signed in Brussels in 1874, the First International Peace Conference in The Hague in 1899, and the Second International Peace Conference in The Hague in 1907 all reached conclusions on specific prohibitions of poison weapons.

However, the limited use of biological warfare weapons during the First World War and the widespread use of chemical agents, paved the way for the subsequent negotiation of a more explicit prohibition that became the 1925 *Geneva Protocol for the Prohibition of the Use in War of Asphyxiating, Poisonous or other Gases, and of Bacteriological Methods of Warfare*. The Protocol as it applies to both chemical and biological warfare is worded as follows:[4]

> *The Undersigned Plenipotentiaries*, in the name of their respective Governments,
> *Whereas* the use in war of asphyxiating, poisonous or other gases, and of all analogous liquids, materials or devices, has been justly condemned by the general opinion of the civilized world; and

> *Whereas* the prohibition of such use has been declared in the Treaties to which the majority of the Powers of the world are Parties; and
> *To the end* that this prohibition shall be universally accepted as a part of International Law, binding alike the conscience and the practice of nations;
> *Declare*:
> That the High Contracting Parties, so far as they are not already Parties to Treaties prohibiting such use, accept this prohibition, agree to extend this prohibition to the use of bacteriological methods of warfare and agree to be bound as between themselves according to the terms of this declaration ...

As it applies to BW against humans, animals and plants, the Protocol represents an extremely general ban on these types of warfare. The Protocol also has a number of weaknesses. As Dando[5] has pointed out, the prohibition essentially represented a no-first-use agreement between states. A strict interpretation of the Protocol finds that the Protocol applies only to war and then only to war between states. Additionally, it applied only to States Parties to the Protocol. The threat of use of biological weapons is not banned by the Protocol and the Protocol contained no provisions for monitoring of states compliance. Again according to Dando:[6]

> in particular, it did not prevent biological warfare being researched or preparations being made for its use – at least in retaliation. Thus the offensive biological research and development carried out by states in the middle years of this century was not banned by the 1925 Protocol.

A further problem with the Protocol in relation to certain types of chemical weapons did not emerge until the 1970s. According to Boserup:[7]

> The wording of the Geneva Protocol does not make it readily apparent whether that instrument was meant to prohibit the use in war of irritant agent weapons and herbicides, but, until the use of these means of warfare began in Viet-Nam, this possible ambiguity had not given rise to serious dispute over the interpretation of the Protocol.

The inclusion of reservations to the Protocol and the failure of some states to ratify weakened the prohibition further. According to Geissler *et al.*,[8] for example:

> By 1939 the major powers (except Japan and the United States) had ratified the Geneva Protocol: Germany, Italy and Poland without reservation: [and] France, the UK and the USSR with stipulations reserving the right of retaliation.

It is important to note that the US did not ratify the Protocol until 10 April 1975.[9]

The latter half of the 20th century saw an initiative to negotiate a more comprehensive prohibition against the use of BW weapons. The BTWC was signed at London, Washington and Moscow on 10 April 1972, with the former Union of Soviet Socialist Republics (USSR), the United Kingdom of Great Britain and Northern Ireland, and United States of America as Co-Depositary Governments. The Convention entered into force on 26 March 1975, and outlaws and entire class of weaponry. In so doing the Convention bans the deliberate initiation of plant disease epidemics as a form of biological warfare, thus reaffirming the commitment of States Parties to the Geneva Protocol of 1925.[10] Like the Geneva Protocol, the BTWC contained no provisions for verifying the compliance of States Parties with their obligations under the Convention. Since 1994 an initiative has been underway to strengthen the Convention with negotiations between States Parties being conducted at the United Nations in Geneva. While a detailed account of the development of the Convention, and the subsequent negotiation of a legally binding compliance and verification Protocol[11] is outside of the purview of this study, the following discussion will consider the relevance of plant pathogens in relation to the above mentioned negotiations. The negotiations to strengthen the 1972 BTWC offer an insight into the extent to which anti-crop BW is regarded as a threat to peace and international security.

The BTWC

The BTWC has a preamble and 15 articles. The following paragraphs give an indication of some of the key points of the Convention. Articles I and II deal with the scope and the prohibition of the Convention. Article I states that:

> Each State Party to this Convention undertakes never in any circumstances to develop, produce, stockpile or otherwise acquire or retain: (1) Microbial or other biological agents, or toxins whatever their origin or method of production, of types and in quantities that have

> no justification for prophylactic, protective or other peaceful purposes; [and] (2) Weapons, equipment or means of delivery designed to use such agents or toxins for hostile purposes or in armed conflict.

Article I of the BTWC contains what is referred to a 'General Purpose Criterion'[12] which represents a sweeping prohibition banning the development, production, stockpiling, acquisition, and retention of agents and weapons for offensive purposes. It will be clear from the outset that Iraq's actions in acquiring offensive BW weapons were in contravention of the General Purpose Criterion embodied in Article I of the 1972 BTWC. The respective Articles of the BTWC are set out in Table 4.1.

Article XIII states the duration and the rights of the State Party to withdraw from the Convention. *Article XIV* dealt with signatures, the process of ratification, and entry into force of the Convention, and *Article XV* dealt with the authentication of the text of the Convention and its deposit in the archives of Depositary Governments.

In accordance with Article XII of the Convention a series of Review Conferences have taken place at five-yearly intervals since its entry into force. To reiterate briefly, the purpose of the review process, in accordance with Article XII of the Convention was, 'to assure that the purposes of the preamble and the provision of the Convention ... are being realized'. During the course of this review process each article of the Convention was therefore considered in turn. Language on the state of realisation of each article of the Convention was agreed by consensus by the Conference and formalised in a so-called Final Document.

The First Review Conference took place in 1981 and Second Review Conference in 1986. The Third Review Conference, in September 1991, mandated an Ad Hoc Group of Governmental Experts (VEREX), to identify, from a scientific and technical viewpoint, potential verification measures for the BTWC. This group met some four times between 1992 and 1993 and recommendations, which evaluated 21 potential verification measures against six criteria, were produced in a final report to all States Parties. In 1994 a Special Conference considered the progress of VEREX and subsequently mandated an Ad Hoc Group[13] to consider 'appropriate measures' to strengthen the BTWC, the objectives:

> including possible verification measures, and draft proposals to strengthen the Convention, [which might] be included...in a legally binding instrument.

Table 4.1 The 1972 BTWC

Article II

Each State Party to this Convention undertakes to destroy, or to divert to peaceful purposes, as soon as possible but not later than nine months after entry into force of the Convention, all agents, toxins, weapons, equipment and means of delivery specified in Article I of the Convention, which are in its possession or under its jurisdiction or control. In implementing the provisions of this article all necessary safety precautions shall be observed to protect population and the environment.

Article III

Each State party to this Convention undertakes not to transfer to any recipient whatsoever, directly or indirectly, and not in any way to assist, encourage, or induce any State, group of States of international organizations to manufacture or otherwise acquire any of the agents, toxins, weapons, equipment or means of delivery specified in Article I of the Convention

Article IV

Each State Party shall take all necessary measures to prohibit and prevent the development, production, stockpiling, acquisition, or retention of the agents, toxins weapons, equipment and means of delivery specified in Article I of the Convention, within the territory of such State, under its jurisdiction or under its control anywhere.

Article V

The States Parties to this Convention undertake to consult one another and to cooperate in solving any problems that may arise in relation to the objective of, or in the application of the provisions of, the Convention.

Article VI

Any State Party to this Convention which finds that any other State Party is acting in breach of obligations deriving from the provisions of the convention may lodge a complaint with the Security Council of the United Nations. Such a complaint should include all possible evidence confirming its validity, as well as a request for its consideration by the Security Council.

Article VII

Each State party to this Convention undertakes to provide or support assistance, in accordance with the United Nations Charter, to any party to the Convention which so requests, if the Security Council decides that such Party has been exposed to danger as a result of violation of the Convention.

Article VIII

Nothing in the Convention shall be interpreted as in any way limiting or detracting from the obligations assumed by any State under the Protocol for the Prohibition of the Use in War of Asphyxiating, Poisonous or Other Gases, and of Bacteriological Methods of Warfare, signed at Geneva on June 17, 1925.

Table 4.1 The 1972 BTWC (*contd.*)

Article IX

Each State party to this Convention affirms the recognized objective of effective prohibition of chemical weapons and, to this end, undertakes to continue negotiations in good faith with a view to reaching early agreement on effective measures for the prohibition of their development, production and stockpiling and for their destruction, and on appropriate measures concerning equipment and means of delivery specifically designed for the production or use of chemical agents for weapons purposes.

The Convention was drafted in such a way that the prohibition would not hamper peaceful uses of biology.

Therefore, *Article X* states that:

The States Parties to this Convention undertake to facilitate, and have the right to participate in, the fullest possible exchange of equipment, materials and scientific and technological information for the use of bacteriological (biological) agents and toxins for peaceful purposes.

Article XI

Any State Party may propose amendments to this convention. Amendments shall enter into force for each State Party accepting the amendments upon their acceptance by a majority of the States Parties to the Convention and thereafter for each remaining State Party on the date of acceptance by it.

Article XII

Five years after the entry into force of this Convention, or earlier if it is requested by a majority of Parties to the Convention by submitting a proposal to this effect to the Depositary Governments, a conference of States Parties to the Convention shall be held at Geneva, Switzerland, to review the operation of the Convention, with a view to assuring that the purpose of the preamble and the provision of the Convention, including the provision concerning negotiations on chemical weapons, are being realized. Such review shall take into account any new scientific and technological developments relevant to the Convention.

Source 'Convention on the Prohibition of the Development, Production and Stockpiling of Bacteriological (Biological) and Toxin Weapons and on Their Destruction', 1972. Available at United Nations Website, Geneva, at *http://www.unog.d disarm.htm.*

The Ad Hoc Group process

Since the first substantive session of the Ad Hoc Group, 10–21 July 1996, and the appointment of the four Friends of the Chair to assist the Chairman on Definition of Terms and Objective Criteria, Confidence-Building Measures and Transparency Measures, Measures to Promote Compliance, and Measures Related to Article X, significant progress was made with regard to the consideration of appropriate measures to strengthen the BTWC. In accordance with the mandate which established the Ad Hoc Group, the Friends of the Chair therefore considered the following:[14]

> Definition of terms and objective criteria, such as lists of bacteriological (biological) agents and toxins, their threshold quantities, as well as equipment and types of activities, where relevant for specific measures designed to strengthen the Convention;
>
> The incorporation of existing and further enhanced confidence building and transparency measures, as appropriate, into the regime;
>
> A system of measures to promote compliance with the Convention, including, as appropriate, measures identified, examined and evaluated in the VEREX Report. Such measures should apply to all relevant facilities and activities, be reliable, cost effective, non-discriminatory and as non-intrusive as possible, consistent with the effective implementation of the system and should not lead to abuse;

and also,

> Specific measures designed to ensure effective and full implementation of Article X, which also avoid any restriction incompatible with the obligations undertaken under the Convention, noting that the provision of the Convention should not be used to impose restriction and/or limitations on the transfer for purposes consistent with the objectives and the provisions of the Convention of scientific knowledge, technology, equipment and materials.

The Ad Hoc Group process continued to report through the four Friends of the Chair. This process was expanded in the latter half of 1997 with four further Friends of the Chair appointed to consider additional measures to strengthen the BTWC.[15] Produced as a series of Working Papers, the recommendations of the respective Friends of the Chair were attached to so-called Procedural Reports at the end of each session of the Ad Hoc Group. The language developed by the Friends of the Chair during the course of the Ad Hoc Group process was subsequently integrated into the first draft of a Protocol to strengthen the BTWC and produced in the form of a so-called 'Rolling Text' in June 1997. While the first Rolling Text contained language for only 2 out of some 20 Articles proposed, by the end of the July 1997 session of the Ad Hoc Group, the Rolling Text contained language for 18 out of the 23 Articles, and at the close of the Eighth Session of the Ad Hoc Group in September 1997, only 2 of the Articles proposed contained no language.

Since the above date considerable progress has been made with regard to the negotiation of the compliance and verification Protocol. Requirements for declarations, procedures for visits to facilities, provisions for investigations together with safeguards for confidential information have emerged from the negotiations as central and essential elements of the Protocol.

As the negotiations approach the end-game there exists the real possibility that agreement on the Protocol will be reached over the next 12 to 18 month period (starting in August 1999) although the final details have yet to be negotiated.[16]

One aspect of the Ad Hoc Group process described above, however, was of particular significance to strengthening the BTWC in relation to the issue of plant disease as a form of biological warfare. We must therefore restrict our attention in the discussion that follows to the element of the Ad Hoc Group mandate that considered appropriate measures to strengthen the Convention in relation to the Definition of Terms and Objective Criteria.

Plant pathogens and the BTWC

During the course of the Fourth Session of the Ad Hoc Group, recommendations put forward by the Friend of the Chair concerned with the Definition of Terms and Objective Criteria produced a list of pathogens of importance to the BTWC. The listing of pathogens included pathogens that affect humans, animals, and plants, and had first emerged during the Second Session of the Ad Hoc Group process.[17] As part of the Ad Hoc Group process, a list of plant pathogens of importance to the BTWC was evaluated against criteria that had been drawn up during the course of its Fourth Session. The listing of plant pathogens important for the BWC was reproduced at the Sixth Session of the Ad Hoc Group, 3–21 March 1997, subsequently expanded at the Eighth Session of the Ad Hoc Group in September 1997, and included in the Annex to the second draft of the Rolling Text.

The identification of particular pathogens during the course of the process to negotiate a protocol was done without prejudice to the agreement contained in the statement of the Final Declaration of the Fourth Review Conference in 1996 which reaffirmed that the BTWC:[18]

> unequivocally covers all microbial or other biological agents or toxins, naturally or artificially created or altered, as well as their components, whatever their origin or method of production, of types and in quantities that have no justification for prophylactic, protective or other peaceful purposes.

According to the Annex of the 'Rolling Text', such pathogens were the subject of:

> further consideration with a view to developing a future list or lists of bacteriological (biological) agents and toxins, where relevant, for specific measures designed to strengthen the Convention.

The content of the Ad Hoc Group Working Papers, according to Pearson:[19]

> reflect[ed] discussions that [had] taken place yet [were] without prejudice to the positions of delegations on the issues under consideration in the Ad Hoc Group and [did] not imply agreement on the scope or content of the papers.

The purpose of this initiative, however, was an attempt to identify a mechanism in the Protocol by which compliance with the BTWC could be verified. In this connection, according to Toth, the list might also aid participating states in the preparation of declarations which include information relating to which agents, activities, and facilities are of relevance to the Protocol for declaration purposes.[20] The final details, however, have yet to be negotiated.

Plant pathogens important for the BTWC

At the Sixth Session of the Ad Hoc Group, 3–21 March 1997, a Working Paper[21] produced by South Africa discussed a number of plant pathogens in relation to the following criteria:

1. Agents known to have been developed, produced or used as weapons.
2. Agents which have severe socio-economic and/or significant adverse human health impacts, due to their effect on staple crops, to be evaluated against a combination of the following criteria.
   a) Ease of dissemination (wind, insects, water, etc.);

b) Short incubation period and/or difficult to diagnose/identify at an early stage;
c) Ease of production;
d) Stability in the environment;
e) Lack of availability of cost-effective protection/treatment;
f) Low infective dose;
g) High infectivity;
h) Short life cycle

The South African Working Paper identified the following 10 plant pathogens, and the diseases with which they are associated, out of a possible 20 plant pathogenic agents, as potential anti-plant BW agents. Based on what the Working Paper refers to as 'internationally accepted CMI (Commonwealth Mycological Institute of the Commonwealth Agricultural Bureaux International) descriptions', the Working Paper included a summary of the characteristics of plant pathogens in relation to the following criteria: distribution; transmission; control; environmental stability; ease of production; and, BW potential. The plant pathogens and a summary of their characteristics are reproduced in full in Table 4.2.

While the criteria against which the pathogens were evaluated did not change, the Annex to the Rolling Text[22] produced at the end of the Eighth Session of the Ad Hoc Group in September 1997, had seen the number of plant pathogens listed increase to some 21 pathogens.

The pathogens identified in the Annex to the second draft of the Rolling Text included the following:

[Citrus greening disease bacteria]
Colletotrichum coffeanum var. Virulans
[*Chochliobolus miyabeanus*]
[*Dothistroma pini* (Scirrhia pini)]
Erwinia amylovora
[*Microcyclus ulei*]
[*Phytophthora infestans*]
Pseudomonas solanacearum
[*Puccinia erianthi*]
[*Puccinia graminis*]
Puccinia striiformiis (*Puccinia glumarum*)
Piricularia oryzae
[Sugar cane Fiji disease virus]
[*Tilletia indica*]

Table 4.2 Plant pathogens important to the BTWC

Name of pathogen	*Disease caused*	*Distribution*	*Transmission*	*Control*	*Environmental stability*	*Ease of production*	*BW potential*
1. *Colletotrichum coffeanum var virulans* [coffee berry disease]	Causes coffee berry disease. Can be very destructive in terms of yield loss and seedling death of this non-staple food crop but does not kill mature plants. Different races have not yet been recorded.	Central and southern Africa.	Seed borne, rain splash, passive vectors such as man, birds and machinery.	Fungicide sprays are not effective. Chemical seed treatment not yet successfully developed. Resistant varieties are available.	Can survive as latent infection. Conidiospores have a short life but conidia can survive more that a year on plant debris	Can be mass produced on artificial substrate but is notoriously unstable under these condition and loses its pathogenicity rapidly.	Not a staple food and thus not regarded as important but may cause serious world wide economic problems.
2. *Dothistroma pini* (Scirrhia pini) (CMI 368) [blight of pines]	Dothistroma blight of pines can be highly destructive depending on the frequency of infection.	Europe, Asia, Africa, North and South America. Different races have not been recorded.	Seed borne, wind, clouds may carry spore inoculum.	Resistant pine species are available. Non-systemic fungicide spray show some activity but are not practical and economically viable.	Inoculum viability debris limited to 2–6 months.	Mass production of the pathogen is easily done on artificial substraits.	Is good although pine is not a staple food it is of strategic [significance?]

Table 4.2 Plant pathogens important for the BTWC (*contd*)

Name of pathogen	*Disease caused*	*Distribution*	*Transmission*	*Control*	*Environmental stability*	*Ease of production*	*BW potential*
3. *Erwinia amyovora* (CMI 44) [fire blight of apple, pear, quince and related species]	Fire blight of apple, pear, quince and related species is very destructive. Not yet recorded in South Africa.	North America, Central America, New Zealand, Japan, China, Europe, North Africa.	Water, vegetative material, insects	Eradicate infected material. Chemical and antibiotic sprays not very successful. The bacteria is not stable in the environment outside its host material. This pathogen can easily be produced in commercial fermenters.			Good.
4. *Pseudomona solanacearum* (CMI 15) [wilt associated with numerous hosts particularly potato, tomato and tobacco]	Potato, tomato and tobacco wilt; slime disease, Granville wilt; bacterial ring disease, Moko disease of banana are some of the most devastating diseases caused	Tropical, subtropial and warm temperate parts of: Asia, Africa, Australasia, Europe, West Indies, North and Central	Infected material, contaminated soil, water, implements	No effective chemical treatments available. Resistant cultivars of varieties but new races develop continuously.			Excellent.

Table 4.2 Plant pathogens important for the BTWC (*contd*)

Name of pathogen	*Disease caused*	*Distribution*	*Transmission*	*Control*	*Environmental stability*	*Ease of production*	*BW potential*
	by this bacterium which attack numerous hosts of Solanaceae, Musaceae, Compositae, Fabacea, etc. Different races of the bacterium occurs that combined with its broad host range make breeding for residence difficult.	America.		The bacterium is stable in soil and host tissue. Spores are not produced and vegetative unprotected cells have limited life span. Easily produced in relatively simple ferments.			
5. *Pyricularia [Piricularia] orzae* (CMI 169) [rice blast disease]	Blast disease of rice can be very destructive (90%) on this staple food. With its many races (219) and broad host spectrum, breeding for resistance is complex. The fungus needs high temperature and humidity for infection.	Widespread: Africa, Asia, Australasia, Europe, North America, South America, Central America, West Indies.	Wind.	Resistant cultivars, sprays of environmentally harmful fungicides can be effective.	Stable, overwinters on straw and debris from reinfection takes place. Can easily be mass produced		Good

Table 4.2 Plant pathogens important for the BTWC (*contd*)

Name of pathogen	*Disease caused*	*Distribution*	*Transmission*	*Control*	*Environmental stability*	*Ease of production*	*BW potential*
6. *Ustilago Maydis* (CMI 79) [maize smut, blister smut and common smut]	Maize smut, Common smut, Blister smut can cause appreciable losses (10–17%). In addition, the spores can induce allergic reaction in man and may be toxic to animals and man. More than 500 races have been noted complicating the search for resistance.	Worldwide where maize (corn) is grown except New Zealand	Wind, seed surface borne, contaminated soil.	Heat or chemical seed treatment but this is useless where soil is contaminated. Possibly resistant cultivars.	Environmental stability is excellent. Spores remained viable after 8 years in dry soil. Can be mass produced on artificial substrates		Good.
7. *Xanthomonas albilineans (CMI 18)*	This bacterium causes leaf scald on sugarcane where [it] can become highly destructive. It has a wide host range and can occur on maize and a number of grass species. The large number of	Africa, Central and South America, Asia, Australasia.	Infected sets, Aerial dispersal, Insects, Rodents.	Heat treatment of sets, resistant varieties. No chemical treatment available.	The bacterium does not produce resistant spores. Disease may remain dormant as systemic infection until environmental conditions favours symptom expression. The bacterium can		Good.

Table 4.2 Plant pathogens important for the BTWC (*contd*)

Name of pathogen	*Disease caused*	*Distribution*	*Transmission*	*Control*	*Environmental stability*	*Ease of production*	*BW potential*
	races complicates breeding for resistance.				easily be mass produced in simple commercial fermenters.		
8. *Xanthomanas compestris pv. oryzae* (CMI 239)	The broad host range bacterium causes bacterial blight of rice and Kresek disease of rice. Kresek is caused by the systemic infection in the tropics and is extremely destructive. Differences in pathogenicity between isolates have been reported but there are no differential varietal reaction to complicate breeding for resistance.	Asia, Africa, South America, Mexico, Korea, Taiwan, Indonesia	Wind, Rain, Flood, vegetative material, Seed borne.	Chemical seed treatment, Resistant cultivars, Elimination of volunteers. Chemical spray not successful	Does not produce resistant or hardy spores. Overwinters on volunteers or in weed shizosphere. Survival on debris seem limited. Can be easily mass produced in simple commercial fermenters.		Medium to good. Candidate for genetic manipulation.

Table 4.2 Plant pathogens important for the BTWC (*contd*)

Name of pathogen	*Disease caused*	*Distribution*	*Transmission*	*Control*	*Environmental stability*	*Ease of production*	*BW potential*
9. *Tilletia tritici* [cover smut, stinking smut and common bunt of wheat]	Cover smut, stinking smut, Common bunt of wheat is caused by this broad host range fungus pathogen which has a single host life cycle. The fungus attacks the inflorescence [flower] replacing the kernels with bunt balls of black teliospores. The disease is regarded as very important, it suppresses yields and lowers the quality and smelly trimethylamine is produced while the spores may ignite and cause an explosion during harvesting.	Worldwide.	Seed surface borne, Wind, contaminated soil.	Resistant cultivars – they are short lived because new races continuously develop. Chemical seed treatment.	Teliospores can survive up to 2 years in soil. Production of this obligate parasite needs live hosts but as vast numbers of spores can be harvested, mass production in not impossible.		Good. Could possibly be enhanced by genetic manipulation.

Table 4.2 Plant pathogens important for the BTWC (*contd*)

Name of pathogen	*Disease caused*	*Distribution*	*Transmission*	*Control*	*Environmental stability*	*Ease of production*	*BW potential*
10. *Sclerotinia Sclerotorium* (CMI 513) [cottony soft rot and white mould of vegetables, beans, sunflower, groundnuts and soya beans]	This plurivourous fungus causes cottony soft rot, white mold, and watery soft rot on a broad host spectrum such as vegetables, beans, sunflower, groundnuts soya bean and many others except cereals and woody plants. The fungus can attack any above ground parts at any development stage and is extremely destructive under cooler moist conditions as found under irrigation.	Worldwide.	Airborne ascospores, Seed infected with mycelium or contaminated with sclerotia (survival structure).				High. Good candidate for genetic manipulation to broaden its temperature spectrum.

Ustilago maydis
Xanthomonas albilineans
Xanthomonas campestris pv citri
Xanthomonas campestris pv oryzae
[*Sclerotinia sclerotiorum*]
[*Thrips palmi Karny*
Frankliniella occidentalis]

The 'agents' identified by the Ad Hoc Group Working Paper and the organisms included in the Annex to the second draft of the Rolling Text include plant pathogens that have been responsible for causing crop losses of considerable significance. The criteria against which they were identified, however, is of particular significance as it represents an important acknowledgement that the utility of crop disease epidemics as BW has been known for some time. Of the agents included in the list compiled during the course of the Ad Hoc Group process this book will show that *Puccinia graminis* (the causal agent of wheat rust), and *Piricularia oryzae* (the causal agent of rice blast), were singled out as agents of particular importance to offensive anti-crop BW developments in the United States, and that research and development on both *Puccinia graminis*, and *Piricularia oryzae* resulted in the standardisation of such agents for military use. Furthermore, the book will also show that attention was given to the consideration *inter alia* of further anti-rice agents, such as *Helminthosporium oryzae* (the causal agent of brown spot of rice), and a further agent involving *Phytophthora infestans* (the causal agent of potato blight).

The potential for such agents to be candidates for genetic manipulation is also a matter that was a cause for concern among those involved in identifying plant pathogens of importance to the BTWC.

While the heavily bracketed text above indicates that a considerable amount of work was still required before the Ad Hoc Group could agree on a list of plant pathogens that may be included in a text, or an annex to, a legally binding instrument, the list contained two notable inclusions which signified a considerable deviation from the list identified by the Sixth Session of the Ad Hoc Group above.

Some additional agents under consideration during the Ad Hoc Group process

During the course of the Seventh Session of the Ad Hoc Group, Cuba submitted a Working Paper entitled '*Unlisted Pathogens Relevant to the*

Convention'.[23] The contents of the Working Paper are reproduced in some detail in the paragraphs that follow. In its introductory remarks the Working Paper commented with regard to the following:

> Cuba finds it appropriate to include in the list of Phytophogens the *Thrips palmi Karny* and the *Frankliniella occidentalis*, agents whose description and assessment with regard to the following selection criteria is the following:

The working paper went on to evaluate the above pathogens against the following criteria:

> 1. Agents known to have been bred, produced or used as a weapon.
> 2. Agents causing serious social and economic consequences and considerable harmful repercussion for human health, due to their impact on basic crops, which will be assessed in combination of the following criteria:
>   (a) easily scattered (air, insects, water, etc.)
>   (b) short hatching period and/or difficult early detection and identification;
>   (c) easily bred;
>   (d) stability in the environment;
>   (e) unavailable cost-effective protection or therapy;
>   (f) low infection dose;
>   (g) high infection capacity [and];
>   (h) short life cycle;

A description of the above agents of dissemination put forward by Cuba for inclusion in the list of plant pathogens of importance to the BTWC appears in the Working Paper included *Thrips palmi Karny*:

> The adult *Thrips palmi Karny* is a very small insect, approximately one millimetre long. Its colour ranges from pale yellow to white. Its wings join over its body drawing a dark line. The eyes are black. It characterizes by attacking leaves where they build up colonies with a large number of individuals. The colour of its elongate, slow-moving larvae ranges from white to yellow. It is a polyphagous agent that causes serious economic damage.

Frankliniella occidentalis is described thus:

> It is a yellow, brown or two-coloured insect, measuring 1.7mm. It is a polyphagous agent that causes serious economic damage.

The above agents of dissemination were then considered against the criteria outlined above.

Agents known to have been bred, produced or used as weapons

According to the Working Paper there are no reports in the literature relating to the use of either *Thrips palmi Karny* or *Frankliniella occidentalis* as agents that have been bred, produced or used as a weapon.

Agents causing serious social and economic consequences and considerable harmful repercussions for human health, due to their impact on basic crops

According to the Working Paper, *Thrips palmi Karny* is considered to be an agent that can cause, serious social and economic consequences and considerable harmful repercussion for human health as follows:

> This species has been reported in several areas latitudes striking numerous varieties of plant (polyphagous) which belong to different botanic families. Likewise, it can also feed on a great number of wild plants and cash crops, namely: cotton, tobacco, onion, egg plant, water melon, pumpkin, cucumber, pepper, potato, sweet potato, bean.

With regard to *Frankliniella occidentalis*:

> The damage caused by this insect are of two kinds: direct and indirect.
> Direct damage:
> a. Seepage of sap through the penetration of the insect's stylet, which tears the epidermic and parenchymatous tissues, causing diverse symptoms depending on the damaged organ and the physiological state of the vegetable, as well as the plate-coating of the fruits, falling of petals and growth disturbances. [and]
> b. inoculation of eggs in the tissues, provoking injuries that cause necrosis or warts.
> The indirect damage is related to its capacity to spread [a plant virus].

Easily scattered (air, insects, water, etc.)

Both *Thrips palmi Karny* and *Frankliniella occidentalis* spread easily on air currents, and on contaminated vegetables and the transit of infected material.

Short hatching period and/ or difficult early detection and identification

Early detection and identification of *Thrips palmi Karny* is considered difficult as the insects, 'implant their eggs within the epidermis of the vegetable tissue, preventing first sight detection'. Similar difficulties regarding detection and identification were found to be associated with *Frankliniella occidentalis*. According to the working paper:

> It is hard to identify inasmuch as there is a great number of species within this genus, so the anatomic characters of adult insects are required to accomplish definition of the species.

Easily bred

It was considered that both insects were easily bred under laboratory conditions.

Stability in the environment

Similarly both insects were reported to be stable in the environment, 'primarily in dry and hot regions'.

Unavailable cost-effective protection or therapy

Thrips palmi Karny 'proved a high capacity of resistance to the use of chemical pesticides. At present, the use of biological pesticides has yielded the best results against it. However, it requires a very costly combined treatment'. *Frankliniella occidentalis* also proved to be expensive to control, such methods, 'being hard to implement as it is necessary to develop combined means of control based on the use of chemicals, biological and culture products'.

Low infection dose

With regard to the infectious dose required to achieve contamination, *Thrips palmi Karny* exhibits a 'high capacity of breeding and adapting to a favourable environment [which] allows for the use of low doses'. *Frankliniella occidentalis* exhibited a similar capacity for reproduction requiring only a low infective dose. According to the Working Paper, 'when the temperature surpasses 25 degrees Celsius, a small amount of insects (25 adults) are able to breed a dense population due to their high procreative capacity and cutbacks in their life cycle term'.

High infection capacity

Both insects were considered to have a considerable capacity to cause infection.

Short life cycle

Both insects proved to have only short life cycles, with adult phases of up to 30 days in the case of *Thrips palmi Karny* and up to 120 days in the case of *Frankliniella occidentalis*. However, during the course of such life cycles both insects were thought to be capable of causing considerable damage to crops from as early as the larval stage.

Cuban allegations

The appearance of the Cuban Working Paper cited above came in the wake of allegations which claimed that crop-destroying agents had been deliberately spread by a US aircraft over Cuban territory. The aircraft, which had been deployed by the US State Department on a drug eradication mission in Columbia, was conducting an authorised overflight of Cuba on route from Cocoa Beach, Florida, via Grand Cayman. Following a Note, dated 28 April 1997,[24] to the Secretary-General of the United Nations describing the appearance of *Thrips palmi karny* in Cuba, Cuba wrote to the Russian Federation as a co-depositary state of the BTWC, requesting a consultative meeting to initiate an investigation under the terms adopted at the Third Review Conference to 'consider any problems in relation to the objective of, or in the application of the provisions of, the Convention'. During a formal consultative meeting attended by over half of the States Parties to the BTWC, and 3 Signatory States, a Cuban statement allegedly linking the US aircraft's over-flight of Cuba's Giron air corridor on 21 October 1996 with the appearance of *Thrips palmy* on Cuban potato crops, was followed by a US statement denying the allegation. The respective statements and further documentary evidence were subsequently distributed to all States Party to the BTWC for review. It was then agreed that submissions containing detailed consideration of the allegations would be submitted to the Chairman of the formal consultative meeting by 27 September 1997. By the deadline, the Chairman of the formal consultative meeting was in receipt of submissions from the governments of Australia, Canada, China, Cuba, Denmark, Democratic People's Republic of Korea, Germany, Hungary, Japan, Netherlands, New Zealand and Viet-Nam. In his subsequent report dated 15 December

1999,[25] the Chairman of the formal consultative meeting, Ambassador Ian Soutar, reported to States Parties that:

> due *inter alia* to the technical complexity of the subject and to the passage of time, it has not proved possible to reach a definitive conclusion with regard to the concerns raised by the Government of Cuba ... there has been general agreement throughout the process that the requirements of Article V of the Convention and of the consultative process established by the Third Review Conference have been fulfilled in an impartial and transparent manner ... the experience of conducting this process of consultation had shown the importance of establishing as soon as possible an effective protocol to strengthen the Convention ...

The above discussion shows that the international community continues to exercise itself over problems relating to the hostile uses of biomedical technologies to the extent that considerable work is underway at the United Nations to identify plant pathogens of significance to the BTWC, and that Iraq's efforts to acquire a capability to wage anti-crop BW took place in parallel to the diplomatic initiative to negotiate a legally binding compliance protocol to strengthen the Convention. It is clear from the above discussion that States Parties are concerned about a very wide range of potential anti-crop BW agents, and a wide range of crops that are vulnerable to such agents.

While it is possible that states other than Iraq have investigated the potential for crop warfare, there is only one example of a substantial and mature programme for which we have considerable publicly available documentation. This is the extensive crop warfare research and development programme which formed part of the United States BW programme from the 1940s through to the late 1960s when the US unilaterally renounced its BW activities and supported the development of the subsequent BTWC. As a result of Freedom of Information legislation and some studies published in the open literature, it is possible to a undertake a reasonably thorough analysis of the US programme.

5
The Context of US BW Research and Development

Domestic developments

The US anti-crop BW programme emerged from domestic institutional arrangements that had been established during the late inter-war years, and the mobilisation of the US scientific community in support of the war effort.

Important external dynamics also influenced the process of armament during this period. Collaboration between the allies on matters relating to BW proved to be of considerable significance. In the light of reported enemy interest in biological weapons, US intelligence activities reinforced the perception that there was an urgent need to acquire BW weapons. Each of the above matters will be dealt with in turn in the paragraphs that follow.

Initially, in the task of preparing for war, the US government was constrained in two important respects. Firstly, war preparations began slowly due to the existence of formal neutrality legislation. Congress set about dismantling the legislation which had formalised American neutrality in the first 18 months of war in Europe, with the arms embargo repealed in late 1939, the first peacetime selective service act in American history passed in late 1940, and the Lend–Lease Bill approved in March, 1941.[1] As Jonas has commented, with such measures in place Congress had:[2]

> legalised the sale of implements of war to belligerents, recruited an army far larger than could conceivably be required to ward off invasion, and rendered the ban on loans to belligerents ineffective.

Subsequently with the outbreak of war in Europe and then later in the Pacific, the prevailing isolationist sentiment that had built up in the US during the inter-war years gradually began to subside.

Secondly, preparations for war began slowly due to the government's own failure to realise that science and engineering had a key role in such preparations. Roosevelt had been persuaded of the practicality of the role of science and engineering in producing new military technologies by prominent individuals in the scientific community,[3] but the US had been slow in establishing domestic institutional arrangements to involve science in the pursuit of weapons production. From such arrangements, however, it would be possible to trace the origins of contemporary science policy[4] and according to Buzan, the origins of the post-War US domestic armament process.[5]

Until such arrangements got underway the US Chemical Warfare Service (CWS), which had been created in June 1918, had utilised, and had to some extent monopolised the talents of the American Chemical Society, but in time a major funding competitor to the Chemical Warfare Service emerged in the form of the National Defense Research Committee (NDRC) in June 1940. The NDRC would subsequently be replaced by an organisation known as the Office of Scientific Research and Development (OSRD) which according to Carroll Pursell was 'by universal agreement, the major American science and technology agency of World War II',[6] but even in this initial phase of the NDRCs existence its success in shaping the institutionalisation of scientific mobilisation was considerable. The Committee[7] of the NDRC was staffed by some of the nations' most important scientists such as the chemist, James B. Conant, president of Harvard, and Frank B. Jewett, President of the National Academy of Sciences, the latter an individual with important links with the science and telecommunications industries. Having received Presidential approval by way of Executive Order, the objectives of the NDRC were stated as follows:[8]

> [to]correlate and support scientific research on the mechanisms and devices of warfare ... aid and supplement the experimental and research activities of the War and Navy Departments; and ... conduct research for the creation and improvement of instrumentalities, methods, and materials of warfare. In carrying out its functions, the Committee [of the NDRC] may ... within the limits of appropriations allocated to it ... enter into contracts and agreements with individuals, educational or scientific institutions (including the National Academy of Sciences and the National Research Council) and industrial organisations for studies, experimental investigation, and reports.

Vannevar Bush, as Chairman of the NDRC, had had an important and direct link to Roosevelt. According to Stewart, 'Bush [had] acted as an informal scientific advisor to Roosevelt and operated at all times with the assurance of the President's support.'[9] This position is thought to have afforded Bush considerable influence in shaping the President's thinking in matters relating to the mobilisation of science in pursuit of the war effort. An important characteristic of that relationship, and one that endured for some considerable time, was that Bush had managed to secure Presidential commitment for the funding of the mobilisation of the scientific community while at the same time maintaining an independence that guaranteed, as Dixon has observed, 'that political controls over science would be exercised only at arms length.'[10] This, in part, went some way to explain why Roosevelt appeared not to be aware for some considerable time of the nature and extent of biological warfare research that was actually being conducted under his presidency during the early 1940s.[11]

Bush's NDRC was not the only initiative concerned with the mobilisation of the scientific community in the early 1940s, and in resisting threats posed to the prominence of the NDRD, Bush was required to perform some skilful bureaucratic manoeuvres. One month prior to the establishment of the NDRC, a rival organisation had been brought to the attention of the President. In a letter to Roosevelt, its chairman had argued for the appointment of a 'central non-military (civilian) committee of engineers, chemists, physicists ... for the purposes of considering and stimulating inventions useful for war purposes, and evaluating their potentialities'.[12] Other organisations proposing mobilisation agendas that were similar to that of the NDRC also subsequently emerged, such as: the Defense Commission; the Department of Agriculture; the National Roster for Scientific and Specialised Personnel; a proposal regarding the establishment of an Office of Scientific Liaison; and, mobilisation proposal put forward by the War Production Board. NDRC prominence prevailed, however, in spite of the emergence of such initiatives.

The most serious threat to Bush's NDRC hegemony over the mobilisation of science in support of the war effort came from the Legislature in the form of Senate Bill 2721. Under the auspices of the Senate Committee on Military Affairs, the Subcommittee on War Mobilisation sought to gather evidence relating to the extent to which the utilisation of the nations scientific and technical resources had progressed. The Bill's subsequent implementation would have allowed for a greater degree of political intervention in the realm of science, placing considerable powers in the hands of the federal bureaucracy. Strenuously opposed to political intervention in the affairs of science, Bush had skilfully involved himself in subsequent revision of the Bill and its implementation had been largely overtaken by events in the latter half of 1944.

By the time that the NDRC had been superseded by the OSRD, Bush was assigning research without consultation with the National Academy of Science (NAS) and the National Research Council (NRC). Prior to the establishment of the NDRC, the NAS and the NRC had been the countries two foremost scientific establishments.

Pursell outlines the significance of the role of the OSRD during the course of this period as follows:[13]

> Throughout the preparedness and wartime period, the NDRC/OSRD was able to carve out for itself the major responsibility for research and development on new weapons, and to defend that responsibility against all efforts either to dilute it with tasks viewed as extraneous or submerge it within some larger, more centralised entity. In part, this defence of its position grew out of a natural bureaucratic effort to protect one's territory. In part, it grew out of an honest conviction that the agency's job was the most important, and that it was doing that job in the best possible manner.

The wider significance of such developments are summarised by Pursell as follows:[14]

> As such, the OSRD, was not only a success within the narrow research limits it set for itself, it was equally successful in perpetuating the scientific leadership which, coming out of the Depression of the 1930s, built the scientific establishment of the Cold War period.

OSRD/NDRC and BW

Early developments relating to the domestic origins of BW research and development in the United States are numerous and complex, and the role of Bush and the OSRD in organising science in support of the war effort in the civilian sector is in evidence throughout this early period. The OSRD also liased with military agencies involved in BW research during the course of this period. Although the CWS had regarded BW to be within its sphere of responsibility since 1924, conducting periodic appraisals of what was referred to then as 'bacteriological' warfare up to the late 1930s, and in spite of military assessments emerging in 1937 that regarded the US to be vulnerable to enemy BW attack, strong sentiments proclaiming BW to be an impractical weapon of war[15] prevailed in the US until the early 1940s.

The military sector began to pay increased attention to matters relating to BW in late 1939 and 1940 with the Chief of the CWS warning in September 1940 that the possibility that enemy use of bacteria and infected insects was a danger that could not be ignored.[16] Preliminary technical studies were instituted under the auspices of the Medical Research Division of the Technical Service at Edgewood Arsenal and tentative programmes for research were drawn up.[17] Persuaded by military assessments of the feasibility of BW that had emerged during the summer of 1941, Harvey H. Bundy, Special Assistant to the Secretary of War, called a joint civilian/military meeting to consider both its defensive and offensive aspects. Representatives from the committees on Medical Research of the OSRD attended the meeting on 20 August 1941 alongside representatives from the Office of the Surgeon General, the CWS, the National Research Council, and Military Intelligence (G-2). Accordingly, the OSRD recommended to Stimson, Secretary of War, that the National Academy of Sciences investigate the matter further.[18]

Stimson subsequently explained institutional developments relating to BW during this period in a letter to Roosevelt in April 1942:[19]

> Some time ago it was called to my attention through my representative in the Office of Scientific Research and Developments, of which Dr Bush is the Director, that there might be serious dangers to this country from what might be described as 'Biological Warfare'. It seemed well to make a preliminary investigation with great secrecy. Accordingly, I asked Dr. Jewett of the National Academy of Sciences to form a secret committee of eminent biologists to consider the question. Such a committee was appointed, including a number of the most important biologists in the country. This committee has made an extensive study and a very thorough report in which it points out that real danger from biological warfare exists for both human beings and for plant and animal life. Some of the scientists consulted believe that this is a matter for the War Department but the General Staff is of the opinion that a civilian agency is preferable, provided that proper Army and Navy representatives are associated in the work.

The 12-member WBC (War Bureau of Consultants) Committee met for the first time on 18 November 1941 and consisted of the following liaison members: Edwin B. Fred (Chairman), Lt Col. Maurice E. Barker, Lt Col. James H. Defandorf, and 1st Lt Luman F. Ney of the Chemical

Warfare Service, and representatives from Ordnance, the Navy Bureau of Medicine and Surgery, The Surgeon General's Office, the Department of Agriculture and the US Public Health Service. In its report, commissioned two months prior to the attack on Pearl Harbor in December 1941, the WBC Committee's assessment of the feasibility of BW reached Secretary of War Stimson, recommending that offensive and defensive BW measures be formulated. The WBC Committee offered something of a rationale for initiating research and development into BW, as follows:[20]

> [t]he value of biological warfare will be a debatable question until it has been clearly proven or dis-proven by experience. The wide assumption is that any method which appears to offer advantages to a nation at war will be vigorously employed by that nation. There is but one logical course to pursue, namely, to study the possibilities of such warfare from every angle, make every preparation for reducing its effectiveness, and thereby reduce the likelihood of its use. In order to plan such preparation, it is advantageous to take the point of view of the aggressor and to give careful attention to the characteristics which a biologic offensive might have.

The institutional origins of BW research and development in the United States had been concealed from the public gaze in a civilian 'New Deal' welfare agency known as the Federal Security Agency (FSA), a body administered by Paul V. McNutt, which oversaw the Public Health Service and Social Security.[21] The order to Stimson that established BW research under the auspices of a civilian non-research agency came directly from Roosevelt with some evidence suggesting that this development had been sanctioned without the President being fully conversant with Stimson's plan to hide the agency. According to Bernstein, 'After a cabinet meeting on May 15 [1942], Roosevelt admitted he had not yet read the secretary's plan but told him to go ahead with it anyway.'[22]

The thinking behind this curious choice of institutional camouflage is still the subject of speculation. While such security measures would clearly help in preventing the enemy from obtaining intelligence information relating to US BW activities, some commentators also point out that obvious military links with research related to BW, during this formative stage of the US programme, and in the early years of the war, may have risked offending the moral sensibilities of a US public that was thought to be strenuously opposed to the use of poison and disease in warfare.[23] Stimson appeared to be trying to convey the sense of discretion with which the issue of BW was treated during this period in a letter to Roosevelt:[24]

> Biological Warfare is, of course, 'dirty business', but in the light of the committee's report, I think we must be prepared. And the matter must be handled with great secrecy as well as great vigour. ... Entrusting the matter to a civilian agency would help in preventing the public from being unduly exercised over any ideas that the War Department might be contemplating the use of this weaponry offensively.

Under the auspices of the FSA, a civilian advisory committee, known as the War Research Service (WRS) [25] was established in mid-1942 and the WBC committee disbanded. After approaches to a number of prominent individuals in the scientific community had failed,[26] George W. Merck, the president of Merck pharmaceutical company, finally took up the position to head the new group in August 1942. The civilian War Research Service, which consisted of eight members including some drawn from the WBC committee, became the central co-ordinating body for all BW research and development during the course of this period, with Merck acting as special advisor on BW to Stimson. Under conditions of great secrecy, the WRS sought to expand biological warfare research and development. The WRS undertook an on-going appraisal of matters relating to BW reporting to the following governments agencies: the CWS, the Medical Department, the Bureau of Medicine and Surgery of the Navy, the US Public Health Service, the Provost Marshal General's Office, the Assistant Chief of Staff G-2, Office of Naval Intelligence, Office of Strategic Services, the Federal Bureau of Investigation and the Department of Agriculture. A further civilian committee known as the ABC Committee, consisting of nine distinguished academics, acted in an advisory capacity to the newly formed WRS Committee, and in the formative stages of BW research and developments in the US, assumed some of the responsibilities of the now defunct WBC Committee.

While Roosevelt personally sanctioned budgetary increases for the WRS, which rose from $200,000 in 1942, to $460,000 in 1944, Roosevelt's own files contain few references regarding the particulars of research and development on BW during the course of this period. According to Bernstein:[27]

> [i]n February of 1943 McNutt informed President Roosevelt that the last of the WRS's $200,000 was being spent. The president, he said, would have to decide whether to 'go more deeply into two or three ... projects now under way.' By April, with Stimson's approval, McNutt requested another $25,000 for the WRS fiscal 1943 budget and a total of $350,000 for fiscal 1944. Two days later Roosevelt endorsed NcNutt's request with one laconic notation: 'O.K. F.D.R.' The WRS 1994 budget grew again several months later, when Roosevelt expanded it to $460,000.

In spite of the apparent absence of written evidence, which conveyed the particulars of the US BW programme in its early stage to Roosevelt in any detail, it is understood that the President may have been briefed on such matters 'off-the-record', as it were, by the Army Chief of Staff, General George C. Marshall.[28]

WRS work was initially sub-contracted to laboratories in the university sector with some 28 research contracts issued, and gradually, as capital expenditures rose and new facilities came on stream, the CWS took over the lion's share of research and development. Considerable reluctance on the part of the War Department prevented military participation in BW research until as late as November 1942.

A directive from the WRS entitled 'Request for Supplemental Research and Development', 10 December 1942, called on the CWS to assume direct responsibility for the military aspects of the BW programme. In the context of alarming German, Italian and Japanese expansionism, significant increases in defence appropriations were sanctioned and government money began to flow into weapons development programmes, such as the atomic bomb project, and via the Army, into the CWS, and its plethora of chemical and BW research projects. By November 1943 the BW centre that had been established by the CWS at the 500 acre Camp Detrick site was fully operational.

The OSRD assumed responsibility[29] for civilian BW related research and developments and the WRS was disbanded. Expansion with regard to further BW facilities continued with development, testing, and production facilities established at the 2,000 acre installation at Horn Island, Mississippi Sound, the 250 square mile site at Granite Peak near the Dugway Proving Ground, Utah, and the Vigo[30] BW agent production facility, Indiana, which was occupied by the Special Projects Division in the Spring of 1944. CWS appropriations leapt from some $2 million in 1940, to $60 million in 1941.[31] Due to the proximity of shipping, and humans, it was soon found that the installation at Horn Island was unsuitable for large-scale testing of biological weapons agents and only two toxins were subsequently tested at this location with the majority of testing taking place at the Granite Peak facility.

Special Projects Division

As a division of the CWS, the Special Projects Division was established in January 1944 to administer the US BW programme. According to Brophy *et al.*,[32] the Special Projects Division would:

> develop measures for defence and retaliation against BW ... produce or procure the necessary material ... collect and evaluate intelligence on enemy activity ... maintain liaison with other military and civilian organisations concerned with biological warfare here and abroad ... prepare training publications and conduct instruction in biological warfare, and ... supply technical advice to the armed forces.

By August 1945, the Division had in excess of 4,000 BW workers with 396 Army officers and 2,466 enlisted Army personnel, 124 Navy officers, 844 enlisted Navy personnel and some 206 civilians. The comparison between the US BW programme and the UK BW programme which began in 1940 (see Chapter 8) was by this time quite pronounced. According to Brophy et al.:[33]

> By comparison, after four years the British BW group under Dr. Paul Fildes at Porton numbered 45, comprising 15 officers and civilians (including 4 officers supplied by Camp Detrick), 20 enlisted technicians, and 10 female helpers.

Under the auspices of the Special Projects Division, BW research and development in the US was ascribed a broad definition, as:[34]

> the intentional cultivation or production of pathogenic bacteria, fungi, viruses, rickettsia, and their toxic products, as well as certain chemical compounds, for the purpose of producing disease or death in men, animals, or crops.

In his report on US wartime BW activities to Secretary of War Stimson, special consultant for biological warfare George Merck reported that:[35]

> A wide variety of agents pathogenic for man, animals, and plants [were] considered [by Special Projects Division scientists]. Agents selected for exhaustive investigation were made as virulent as possible, produced in specially selected culture media and under optimum conditions for growth, and tested for disease producing power on animals or plants.

In its concluding remarks Merck's report also began to take on a somewhat prophetic tone:[36]

> this type of warfare cannot be discounted. ... It is important to note that, unlike the developments of the atomic bomb and other secret weapons during the war, the development of agents for biological warfare is possible in many countries, large and small, without vast expenditures of money or the construction of huge production facilities. It is clear that the development of biological warfare could very well proceed in many countries, perhaps under the guise of legitimate medical or bacteriological research.

The anti-personnel programme would result in the subsequent standardisation of some eight BW agents and their respective delivery systems. After limited production of anti-personnel BW agents had begun with pilot plants at Detrick producing *botulinum* toxin completed in October 1943, and anthrax and anthrax simulant in March 1944, large-scale production of anti-personnel BW agents would subsequently take place at Pine Bluff Arsenal, which was constructed in the mid-1950s. In addition to viral agents, rickettsial agents, and a facility for the production of insect vectors as agents of dissemination, so to speak, according to one declassified source,[37] between 1954 and 1967, Pine Bluff produced the following anti-personnel BW agents and toxins:

> *Brucella suis, Pasteurella tularensis,* Q fever rickettsia, Venezuelan Equine Encephalomyelitis, *Bacillus anthricis, botulinum* toxin and *Staphylococcal enterotoxin.* Bulk agents and anti-personnel munitions filled with these various agents and toxins were produced and stored ... as a deterrent capability.

The Special Projects Division also conducted a limited anti-animal programme. According to SIPRI, anti-animal BW research was conducted into the operational effectiveness of rinderpest virus, Newcastle disease virus, fowl plague virus, and model viral and bacterial agents.[38]

Anti-Crop BW in the US

The following statement represents an overview of the microbial anti-crop programme conducted at Detrick over a period of some 25 years until Nixon's renunciation of offensive BW research and development in 1969, and the destruction of the stockpile in the early 1970s. According to an official US source:[39]

> [r]esearch included strain selection, evaluation of nutritional requirements, development of optimal growth conditions and harvesting techniques and preparation in the form suitable for dissemination. Extensive field testing was done to assess the effectiveness of agents on crops.

Research and development into the operational effectiveness of anti-crop BW agents would subsequently result in the standardisation of three agents and their respective means of delivery designed to destroy crops of major economic and social significance.

With the institutional origins of the US BW programme in place and the mandate of the Special Projects Division clearly defined, the US worked closely in collaboration with its wartime allies during the course of this period. The Merck Report on wartime US BW activities only hinted at the nature and scope of investigations of relevance to developments in anti-crop BW – investigations which, according to Merck,[40] included: 'Studies of the production and control of certain diseases of plants.' Shortly after the war, however, the details of such investigations could be seen for the first time and with security restrictions relaxed technical papers and reports began to appear in the public domain. One such example has been included as Appendix I in order to allow the reader to develop an appreciation of an actual investigation into anti-crop BW. However, the next section turns to assess the extent to which anti-crop BW activities had taken place in France prior to the German invasion. A similar assessment is then made of the perceived threat posed by Germany and Japan.

6
Aspects of Anti-Crop BW Activity in France, Germany and Japan

Debate concerning both defensive and offensive aspects of BW research and developments in the US was fuelled by perceptions of enemy activity in relation to BW. A series of press articles by British journalist Wickham Steed emerged as early as 1934 claiming that Germany was planning BW attacks on the Paris and London Undergrounds, but these and other stories relating the dangers posed by a variety of hypothetical covert BW scenarios against targets in Europe and the US, were afforded little credence by Washington.[1]

After the outbreak of war, intelligence estimates regarding German and Japanese biological warfare activities increased. In the light of intelligence estimates which postulated that Germany's 'buzz-bombs' could be loaded with BW agents and directed against the UK and its troops in northern France, a certain urgency became associated with allied Anglo-American research into a retaliatory BW capability. Preparations were made[2] for the inoculation of approximately 100,000 troops against botulism in a measure that would have signalled a certain level of allied BW preparedness to the Germans. However, lack of allied preparedness with regard to a BW retaliatory capability – at this stage only a limited number of anti-personnel anthrax bombs had been prepared for testing purposes; and preparations were underway in the UK for the production of some 5 million anti-livestock cattle cakes – would probably have resulted in an Anglo-American chemical response had the allies been attacked with BW agents.[3]

There is considerable evidence that the allies had overestimated the intentions of the Germans with regard to their level of BW preparedness. Although German Army BW workers had conducted research into BW since 1941, it was learnt only after the war that German interest in acquiring a capability to wage BW had not proceeded much further than a preliminary phase. Allied interrogations after the war soon led to the identification of key German BW workers and the subsequent recovery of all documentation pertaining to the German Army BW programme from the files that had been hidden in a monastery vault in south-eastern Bavaria.

However, during the First World War, German agents had conducted covert anti-animal BW sabotage operations against animals, and American shipments of horses to the theatre of war in Europe were subject to considerable disruption. According to one report, 'all horses on one transport from America to England had to be thrown overboard because the animals were diseased with glanders'.[4] In what can be regarded as an early example of anti-crop BW, grain shipments were also subject to sabotage during this period.[5] However, little interest in BW[6] in Germany was shown in the inter-war period and BW was considered to have no future by representatives of German science until as late as 1940.

German CW Army Ordinance inspection of four laboratories at Centre d'Etudes du Bouchet (CEB), in France, resulted in the discovery of French activities relating to BW and in a report submitted to the German Surgeon General, 17 September 1940, BW research scientist Kliewe stated: '[w]e learned for the first time how promising the enemy considered this field'.[7]

France

French activities in the field of BW began in 1921 and in developing an appreciation of French BW research, and Germany's subsequent decision to institute a programme of BW, it is possible to identify three reasonably distinct phases in the French programme to 1940. During the first phase beginning in 1921 and lasting until 1926, reports surrounding German covert BW activities during the First World War were a cause of French anxiety, and considerable effort was expended in evaluating intelligence regarding German progress on BW research and development. Although the French government had expressed an interest in establishing the consequences and implications of BW since December 1921, it was not until December 1922 that a special commission on BW, the Bacteriological Commission, was established by Ministerial decree, and work began on defensive measures and the feasibility of a capability to retaliate in-kind. According to Lepick, the remit of the Bacteriological Commission was as follows: [8]

> to maintain close liaison with the Commission des Etudes et Experiences Chimiques on questions associated with bacteriological

> warfare, to collate all information and documentation liable to have relevance to bacteriological warfare and, finally, to predict those methods which an enemy might employ in a bacteriological conflict and to undertake all studies necessary for the establishment of a defensive organisation and, if required, for a counter-offensive.

French BW workers conducted laboratory research on the utility of agents and their means of dissemination and work was done on the behaviour of microbial clouds and a number of field trails were carried out with some success. The Bacteriological Commission assumed responsibility for the allocation of funding of BW research, and was free to decide upon the direction that research should take.

French signing of the Geneva Protocol in June 1925 ushered in a second phase regarding French activities associated with BW resulting in the considerable curtailment of the programme until 1934. During the course of this period the Bacteriological Commission ceased to function and BW activities were conducted under the auspices of the Commission des Etudes et Experiences Chimiques, with work restricted to basic research and defensive measures.

A third phase began in 1935 and lasted until 1940. Germany had left the League of Nations in October 1933 and in the context of increased tension in Europe, French interest in BW was rekindled. The Bacteriological Commission was revived and met for the second time since 1922 in December 1935, and the French BW programme saw considerable expansion. French activities during the course of this period focussed on developing means with which to retaliate and in order to do this, offensive BW research received official French War Ministry backing in 1937. Allegations regarding German BW activities caused considerably more alarm in France than they did in Washington, and allegations regarding German covert operations on the London Underground and the Paris Metro resulted in trials in Paris confirming the dangers associated with such a form of attack. The Bacteriological Commission was renamed, the Commission de Prophylaxie (Prophylaxis Commission) for security reasons, and BW research subsequently transferred in its entirety to the Poudrerie Nationale du Bouchet. The Prophylaxis Commission met every six months between January 1938 and September 1939, and then every three months from September 1939 to June 1940. With intelligence reports of German activities given high priority, the meetings of the Prophyaxis Commission became a forum where on-going research and developments associated with BW could be evaluated. Although data relating to French BW activities in this section is restricted to a single source, the following information (Table 6.1) is thought to offer an appreciation of the range of BW research and development under consideration by the Prophylactic Commission during the course of this third phase:

Limited Anglo-French collaboration on matters relating to BW had begun in 1940. In this connection, the two countries established communication and intelligence information relating to German activities was exchanged. Information-sharing, however, was limited to defensive measures only and a link was established between Lord Rothschild of the British War Office and French Army Staff. Lord Hankey received a French delegation on matters relating to defence against BW attack in May 1940, and collaboration in this connection was largely overtaken by events and the cessation of the French programme. Although the German invasion had brought an end to BW research in France, during the third phase of the French programme anti-crop research appears to have centred around the possible use of Colorado Beetles, and a plant pathogen (*Phytophthora infestans*), as a possible means of conducting an economic assault against German potato crops. Although such research had clearly not resulted in an operational capability to wage anti-crop BW, French activities will be seen to be consistent with the steps toward acquiring an operational capability in BW as identified by the US Office of Technology Assessment in Table 1 (see Stages 1 to 4).

Germany

On the basis of French research, experimentation and field trials were planned under the supervision of Kliewe and the German programme was expanded and research organised under the auspices of the Blitzarbeiter Committee with separate bodies responsible for conducting defensive BW research against humans, animals and plants. Developments relating to weaponry were assigned to the German Army's Ordinance Department.

German BW preparations, not unlike BW preparations in Allied countries, appear to have been spurred on by intelligence reports of enemy activity in BW. Such reports, collected by both civilian organisations and the German Army, alleged instances of sabotage in German occupied areas, and it is reported that defensive preparations in the field of BW were initiated in response to certain reports. Whilst German interrogation of French BW workers alleged that the most likely method of English BW against the Germans could involve a deadly cocktail of anthrax and mustard gas, there is little evidence in

Table 6.1 BW Research during the third phase of the (French) prophylactic commission 1935–40

Date	*Details of research under consideration*	*Comments on research under consideration*
January 1938	1. Ricin research. 2. Ricin vaccine and serum. Consideration of inert particles as carriers of BW agents. 3. Germicidal paints and coverings.	1. Ricin research considered too dangerous until more progress had been made with vaccine and serum.
June 1938	1. Study of botulinum toxin released in air. 2. Vaccine and serum for ricin continued.	– 2. Danger to water supplies from ricin contamination reported. Increased emphasis upon water purification/monitoring.
May 1939	1. Work on inhalation of ricin toxin. 2. Work of modes of attack (a). projectiles and (b) dispersion in air. 3. Bovine plague and fever research. 4. Decontamination of water supplies. 5. Aerial delivery of Colorado beetles and insects to impair agriculture. 6. Research on fungal plant pathogen (*Phytophthora infestans*).	1. Toxic action by inhalation of ricin toxin reported as promising. 2. Work continued to collaboration in collaboration with dispersion and physiology laboratories at Le Bouchet. 3. Presented problems relating to dispersion. 4. 700 filtration units under construction. 5. Research stressed economic importance of potato crops in Germany. 6. Use of this pathogen reported to have been considered during the First World War, presenting danger to potato plants.
September 1939	1. Contaminated projectiles. 2. Bovine plague. 3. Ricin absorption through lungs.	1. Trials reveal possibility of infecting animals using contaminate projectiles.

Table 6.1 BW Research during the third phase of the (French) prophylactic commission 1935–40 (*contd*)

Date	*Details of research under consideration*	*Comments on research under consideration*
	4. Botulinum toxin. 5. Abrin research. 6. Colorado beetle.	2. Bovine plague very contagious and animals easy to infect. 3. Lungs not a good absorption route for ricin. 4. The effects of munition explosion on botulinum toxin. 5. Toxic effects of Abrin on the conjunctive and opthalmia requested. 6. Field trials and breeding programme instituted to establish effects of altitude on beetles, effects of release in flights, and mode of dispersion.
December 1939	1. Aircraft canister and grenade trials. 2. Bovine plague. 3. Anthrax.	1. Contamination properties of aerosol spores effective. 2. Research suggested using aerosol of bovine plague. 3. Anti-anthrax serum requested.
January 1940	1. Bovine plague. 2. Combined munitions	1. Aerosol of bovine plague deemed to be very feasible. 2. Certain gases deemed to have positive role in the development of latent anthrax infection.
April 1940	1. Means of dissemination.	1. All research subordinated to development of aircraft bomb and artillery shell. Field trials at Somme estuary against animals.

the public domain relating to corresponding changes in German BW preparations to counter such an offensive in-kind, but reports disseminated in April 1942 that England was preparing to attack German food crops resulted in considerable defensive activity. In April 1942, in response to a report which warned that England had received a consignment of Colorado beetles for dissemination over Germany, German High Command noted that, 'information regarding the defence against potato beetles was urgently required'.[9] Further reports distributed in January 1943 revealed that experimentation into the effects of anthrax and foot and mouth disease had been conducted at Edgewood BW facility in the US, and strict controls through prophylactic immunisation of cattle remained in place throughout this period.

Respective German BW agencies instituted a limited testing programme. Testing of anti-personnel agents conducted under the supervision of Kliewe at Munsterlager (a testing station) in July 1942, initially involved free-fall bombs (designated KC 250 II Gb) filled with agent simulant (*Bacilli prodigiosus*) and subsequent testing included the use of fragmentation bombs (Zerlegerbomben, designated, 2AB 50 m) filled with agent simulant and straw payload (the latter presumably to prevent simulant being carried off by wind currents that had been reported to be a problem in relation to previous testing). The testing was conducted using a Dornier 17 aircraft and in spite of technical problems with the fuses associated with the bombs, a report submitted in May 1944 revealed that the, 'KC 50 AB ... had] shown good results'.[10]

Further testing with foot and mouth disease took place in the summer of 1942 and 1943, and while there appears to be no available information relating to the success of this aspect of the German experimental programme, one such test is reported to have involved the use of a crude aircraft spray device.

In comparison to testing of agents and ordinance against man and animals considerably more attention was given during the course of German activities to BW against crops. An ALSOS Mission[11] Intelligence Report relating to the Agricultural Section of the German programme reported as follows:[12]

> There were probably more plans and ideas considered by the workers in the plant section than in any of the other three sections. There is a repeated emphasis on the possible use of different agents for attack against England and in one case America is specifically mentioned.

Although the data in the following table (Table 6.2) is drawn from only a single source,[13] it is thought to offer an accurate appreciation of some of the plant infections and plant parasites considered by German BW workers:

Table 6.2 Plant infections and plant parasites considered by Germany

German name in documents	*Probable identity*	*Year*	*Details of references of work planned or executed*
1. Kartoffelkafer	Potato beetle colorado beetle	1941 1942	Mentioned By Kliewe Subject of Intelligence reports from US
2. Rapaglanzkafer	Rape seed beetle	1943	Mentioned as a possible BW weapons
3. Rubenaaskafer	Turnip rot beetle	–	(In some reports no clear distinction made as to which pest, reference is made.)
4. Rubernrusselkafer	Turnip leaf bug	1944	Use as a complement to the use of potato beetles. Later reports work in progress.
5. Rubenblattwanze	Turnip leaf Bug	1944	Experiments planned a B. Field East testing station.
6. Kornkafer (Calandra)	Corn beetle	1943	No interest as supplies are not available.
7. Getreidewanze	Grain bugs	1943	These are of lesser importance than pasture gnats and turnip parasites.
8. Weizenschadling	Wheat 'blight'	1943	Parasites are to be imported from Rumania and Turkey to be used in cultivation experiments.
9. Weizenalchen (*Tylenchue tritici*)	Worm infestation of wheat	1941	Possible BW weapons.
10. Spargelkafer	Asparagus beetle	1943	Experiments impossible as the beetles are not available.
11. Japnische Kafer	Omniverous beetle from Japan	1943	Experiments are planned to see if these beetles will withstand German climate. [Supplies] of these beetles are expected to arrive from Japan within four weeks.

Table 6.2 Plant Infections and plant parasites considered by Germany (*contd*)

German name in documents	*Probable identity*	*Year*	*Details of references of work planned or executed*
12. Wiesenschnaken	Pasture gnats	1943	March. Worth considering the use of these but no experience is available.
13. Grassule (Schmetterling)	Butterfly	1943	September. Plans to experiment with these. They are said to have destroyed 43,000 hectares of pasture in one year. May 1944. Experiments have given no clear-cut results as yet.
14. Weidenblattkafer	Willow leaf beetle	1943	July. Experiments impossible as supplies are not available.
15. Nonne	Night moths	1943	March. Suggestion that these be used in America as they are not numerous over there.
16. Riefenblattweape (Diprion. RNI.L)	Pine leaf wasps	1944	Experiments planned for B. Field East testing station.
17. *Tilletia tritice* (Ustilaginae)	Wheat blight	1941	Mentioned as possibility.
18. Getreide rost (Gelb, Braun, Schwartz)	Grain 'rust' yellow, brown, black	1943	Valuable. All varieties must be used. Mentioned to Himler as possibility. Techniques of distribution and production [discussed]. June. Research limited in favour of this work.
19. *Puccinia Glumarum*	Rusts	1941	Possible BW weapon.
20. *Septora tritici*	Wheat pest	1944	Experiments reported as negative so far. June. Will not be used against Germany but possibly against wheat in SE Europe.

Table 6.2 Plant Infections and plant parasites considered by Germany (*contd*)

German name in documents	*Probable identity*	*Year*	*Details of references of work planned or executed*
21. Cercospore	Turnip fungus	1944	June. Reported to have been cultivated on artificial media. Collection of bulk material from dried infected turnip leaves is possible. Problem of distribution not solved. Experiments at B. Field East testing station are planned.
22. Kartoffelkrautund Knollen Faule (*Phytophora Infestans*)	Potato rot	1941	Pointed out dependence on wet weather for its spread and development.
23. Unkraut	Weeds	1943	Little is known about their application. 1944. May. Experiments in early stages. 1944. June. Experiments in progress at Kaiser Wilhelm Institiute for Horticulture.

A more detailed account of German offensive preparations, however, was included in the ALSOS Intelligence report in relation to potato beetles and grain rusts. By July 1943 German BW workers had formulated vulnerability assessments of agricultural targets on the English mainland. A committee meeting in March 1943 revealed the following:[14]

> [i]n considering the use of plant parasites, foremost would be the use of potato beetles against England where the defence against them has not been organised ... England probably has about 400,000 hectares of potato fields for whose destruction 20–40 million beetles would be necessary ... destruction of the potato crop would only bring about a reduction of 12% in the caloric value of England's food supply ... the food supply of the cattle might also be attacked by destroying the turnip crop with the turnip beetle.

In an attempt to obtain the number of Colorado beetles required for an attack on England, a programme involving the large-scale breeding of beetles was instituted in the middle of 1943, and in order to establish the pattern of dissemination of beetles dropped from aircraft field trials were planned for November of that year. However, during field trials conducted in the Speyer region only 57 of the 14,000 beetles dropped from an aircraft were subsequently recovered by German BW workers. Field trials were subsequently continued using wooden models of beetles with improved recovery rates for such models reported to be in the region of 70 per cent over an area radius of some 10 km from the point at which the models had been released. By June 1944, one report appeared to indicate that Germany had acquired an operational capability involving the use of beetles as a means of conducting BW. In spite of reporting the absence of data relating to the effects of extremes of temperature upon beetles, it was reported as follows, that, 'their use is possible at any time'.[15]

Early investigations into the possibility of deploying plant pathogens against enemy food crops revealed that fungus infection of grain might have resulted in a more effective weapon than the grain bug (Getreidewanze) that was also under consideration. A committee meeting in July 1943 revealed, however, that 'several varieties – yellow, brown and black blight – would have to be used in order to ensure a 50% destruction of crops'.[16] Laboratory experimentation had demonstrated various means of dissemination with fungal spores mixed in combination with talcum powder. Limited field testing was conducted into the destructive effectiveness of stalk rot of potatoes (Krautfaule) and *Phytophthora infestans*. Further work with fungal plant pathogens appears to have been discontinued as Germany had no defence against its use.

While some progress had been made with regard to offensive BW preparations in Germany, research and development during its initial stage appears to have taken place in the absence of clear policy directives from German High Command. Reports suggest that it was not until May 1942 that policy directed that, in accordance with Hitler's wishes, German BW research and development would be strictly confined to defensive work only. While German Army High Command orders preventing offensive BW research and development were subsequently repeated again on two separate occasions in 1942, it is reported that Hitler again issued a directive in March 1943 re-affirming the prohibition on offensive BW research and development. However, offensive BW research and development was subsequently permitted in so far as its investigation took place in connection with devising defensive measures, and offensive research appears to have progressed in this context. During a conference at the General Staff of the Supreme German Command Kliewe noted[17] that no meaningful distinction could be made between defensive and offensive biological warfare research at the experimental level:

> Since appropriate protective measures cannot be undertaken without knowing the methods by which bacterial attacks can be brought about, it is necessary that we carry out experiments. You have to attack yourself before you can evaluate protective measures correctly.

According to the ALSOS Intelligence Report, '[w]hatever progress was made in the study of offensive methods resulted from the individual initiative of the workers, continually impeded by the official policy,'.[18] Although limited offensive BW research and development took place in wartime Germany, such activities were not inconsistent with the steps toward acquiring an operational BW capability as identified by the US Office of Technology Assessment in Table 1 (see Stages 1 to 4). However, German adherence to the 1925 Geneva Protocol was maintained throughout the Second World War.[19]

Japan

Reliable allied intelligence on Japanese BW activities emerged only toward the end of the war. By the time US agencies responsible for collating intelligence information under the auspices of the Military Intelligence Services, the Office of Strategic Services, and the FBI had been co-ordinated in 1943, reports had surfaced that on a number of occasions Japanese scientists had attempted to obtain cultures of dangerous biological organisms from laboratories in the US. The allies had also been alerted to a series of detailed allegations of Japanese BW activities in China. Such claims, which centred around Japanese activities in China's Chekiang and Hunan regions, where the outbreak of bubonic plague had resulted in a significant number of deaths in 1940, were largely dismissed by British officials. However, further allegations subsequently prompted condemnation in the US with Roosevelt warning in 1943 that if Japan did not cease, 'retaliation in kind and in full measure would be metered out'.[20] A critical mass of information and allegations relating to Japanese BW activities in China had resulted in allied intelligence being put on full alert in 1944. Only after the war did the full implications of Japanese BW activities come to light.

From literature that has emerged in more recent years into the public domain, the origins of BW research in Japan can be traced back as far as the early to mid-1920s. Interrogation of Japanese prisoners of war (POWs), and re-interrogation of POWs who were found to have had an association with medical activities and especially water purification, subsequently revealed a biological warfare programme of considerable size. While Japanese BW was conducted on a small scale in the mid-1930s, a systematic approach to research and development was established by a decree issued by Emperor Hirohito in 1936, and Japanese BW activities organised under the auspices of Unit 731. This development propelled Major Ishii Shiro,[21] who was by now a long-time campaigner for a major programme of BW in Japan, and Unit 731, from relative obscurity to great national significance.

The Ping Fan Headquarters of Ishii's BW facility was based in a suburb of Harbin, in Japanese occupied Manchuria. Unit 731 employed a workforce of some 3,000 with BW work organised under respective sections dealing with bacteriological research, warfare research and field experiments, water filter production, bacteria mass production and storage, education, supplies, general affairs and clinical diagnosis.

The most notable characteristic of BW research conducted under Ishii's guidance was that an estimated 10,000 humans were sacrificed during the course of experimentation to refine the hostile uses of biomedical technologies.[22] In research designed to establish the optimum infectious dose of agents, and the most efficient route through which infection could be achieved in humans, Japanese BW workers considered a wide variety of infectious diseases. The following list of infectious diseases gives an indication of the scope of agents under consideration: anthrax, botulism, brucellosis, cholera, dysentry, epidemic haemorrhagic fever, fugu (blowfish) toxin, gas gangrene, glanders, influenza, meningococcus, plague, salmonella, smallpox, tetanus, tick encephalitis, tuberculosis, tularemia, typhoid (and paratyphoid), typhus and tsutsugamushi fever.

Problems associated with the dissemination of infectious diseases lead to the study of insect vectors as agents of dissemination. More than a dozen species of fleas that were known to bite man were considered as potential carriers of plague bacillus. Plague infected rats were used to breed fleas and huge numbers of rats and fleas were housed in Unit 731's animal houses. With the matter of protecting the plague cultures from the effects of environment now solved, getting the fleas to the target was overcome with the aid of specially adapted free-fall 25 kg porcelain munitions known as Uji bombs. Testing took place at the isolated Anta test site in Manchuria with the aid of a squadron of aircraft designated the Heibo 8372 Field Aviation Unit. Early studies with artillery proved it unsuitable as a means of dissemination for bacteria. Testing was even conducted to establish the operation effectiveness of feathers dusted with infectious diseases. Hundreds of tests were conducted in order to establish the operational effectiveness of 9 bomb types and their respective payloads. Two further variations of the porcelain Uji type bomb were developed. A glass bomb designation Ga was developed and five iron bombs with designations, I, Ro, Ha, Ni and U. Bombs containing the causal agents of anthrax – regarded as particularly well suited to aerial dissemination in the form of spray due to its spore characteristics that were thought to shield the bacillus from environmental decay – and the causal agents of gas gangrene, and plague, were all tested against live humans. Plague was thought particularly suited to Ishii's BW needs as it was endemic throughout Asia, China (including Manchuria) and India, and the origin of its deliberate initiation could be easily passed off as a natural occurrence. Such agents along-side dysentery, tetanus and typhoid were considered to be most effective on the battle field.

Another notable characteristic of Japanese BW research and development related to the un-scientific approach adopted by the Japanese in some aspects of its work. According to a declassified report on Japans' BW activities:[23]

> Japanese offensive BW was characterised by a curious mixture of foresight, energy, ingenuity, and at the same time, lack of imagination with surprisingly amateurish approaches to some aspects of the work.

Thus important quantitative data from experimentation with cloud chambers used on a limited scale during the course of the British programme and subsequently utilised to a much greater extent during the course of the US programme was not available to the Japanese BW workers.

Covert operations involving the contamination of water supplies and the dissemination of disease infected rats and fleas into densely populated areas were thought to be highly effective potential forms of warfare. Many other methods of dissemination were considered for sabotage during the course of Japanese research and development into offensive anti-personnel BW.

Japanese BW workers conducted research into infections diseases that effect animals under the auspices of the Hippo-Epizootic Unit of the Kwantung Army otherwise known as Unit 100. Established at the same time as Unit 731 in 1936, Unit 100 employed some 700 BW workers. Amongst agents selected to target animals, Unit 100 BW workers conducted bacteriological research into the effects of horse poison using anthrax, potassium cyanide and strychnine. Other agents were also considered for their effects upon sheep and cattle and Unit 100 was capable of producing 500 kg of glanders during the course of one annual production cycle.

Unit 100 BW workers also conduced anti-crop BW research. Much of the arable land occupied by Unit 100 was utilised for the cultivation of basic and exotic plant crops with experimentation being conducted by a squad of 11 of Ishii's workers under the supervision of Dr Yukimasa Yagisawa into the destructive effectiveness of a variety of plant pathogens and chemical herbicides. Williams and Wallace have outlined some of the activities of this section as follows:[24]

> For their target areas, Siberia and the Pacific North-West, this squad selected stinking smut of wheat and nematosis as agents. Yagisawa proposed collecting these agents as by-products of flour-milling operations in Manchuria. For nematosis he discovered that as little as 5 kilograms per acre would ensure complete destruction of susceptible varieties. Experiments showed that most Japanese and Australian varieties were 80 per cent susceptible; Russian varieties 25per cent and American ones somewhere in between.

Under a directive which called for the mass production of red rust germs, Unit 100 was capable of an annual production capacity of 100 kg. Japanese BW workers also conducted research into the destructive effects of mosaic plant diseases as potential anti-crop BW agents. It will be noted that such developments are consistent with the steps that are required in obtaining an operational biological warfare capability as identified by the US Office of Technology Assessment in Table 1 (see Stages 1 to 4).

In the light of allegations that Japanese soldiers and Japanese Army horses had been subject to BW sabotage attacks by Chinese guerrillas, Ishii's BW programme initially procured funding in the mid-1930s under the guise of the pursuit of a capability with which Japan could retaliate in kind. Funding was subsequently obtained from the Japanese War Ministry via the Kwantung Army and Units 731 and 100 are reported to have had a budget of approximately 10 million yen and 2 million yen respectively.[25]

In October 1941, Ishii received a high level commendation for work that is thought to be linked to the discovery and development of the porcelain bomb – an award that was issued in person in by Hideki Tojo Japan's wartime premier. It is understood that this award was presented as confirmation of the success of this technology in a raid on China which is reported to have caused a minor epidemic resulting in the loss of some 24 lives. Massive quantities of combinations of anthrax, cholera, dysentery, para-typhoid, plague and typhoid germs were subsequently used in Japanese BW campaigns against the Chinese nationalists new capital at Chungking, and against the central Chinese province of Chekiang, the latter resulting in the loss of some 2,000 lives. According to Williams and Wallace: [26]

> So great were Chinese losses resulting from Unit 731's activities during the Chekiang Campaign that one authority later described them as 'inestimable'.

It is also understood that Japan's Premier had been party to the nature of experimentation conducted under the supervision of Ishii with reports confirming that Tojo subsequently developed an aversion to motion pictures of human experimentation. It is also known that senior military and political figures were familiar with the activities at Japanese BW facilities and on at least one occasion a member of the Emperor's family visited Unit 731 during the course of its activities in Manchuria.

Perceptions of threat and the accuracy of intelligence

Prior to the war and during the course of its early years, perception of enemy activities relating to BW relied heavily upon the gathering of intelligence assessments. In allied countries such reports and a series of allegations appearing in the popular press were met with varying degrees of anxiety. Assessments of German BW preparedness gave too much credence to the progress that had been made with regard to Germany's attempts to acquire a capability to wage BW and with the benefit of hindsight such assessment could not have foreseen that attempts to acquire a German offensive BW capability would be thwarted by official policy prohibiting such developments.

With regard to intelligence estimates of Japanese BW activities the opposite was true. On a number of occasions the Chinese had made a series of allegations relating to Japanese offensive BW activities but it was not until General Chiang Kai-shek, the then Chinese Nationalist leader, had made a formal approach to the British government in July 1942 that such allegations were treated with any degree of seriousness. Through his Ambassador in London, Dr Wellington Koo detailed allegations of five separate instances of alleged Japanese BW activities in China and these allegations were considered before the Pacific War Council which was attended by both Anthony Eden and Winston Churchill. However, again, the allegations fell on deaf ears and Chinese claims were denied by the allies. According to Williams and Wallace, however, '[s]uch arrogance was misplaced. Koo's report was startlingly accurate'.[27] Allied intelligence was not on put on full alert until as late as 1944 and there remained an absence of reliable intelligence information relating to Japanese BW capabilities until after Japan's surrender and the subsequent interrogation of key Japanese BW workers and prisoners of war.

7
Post-Second World War US Anti-Crop BW

At the announcement of Japan's surrender on 2 September 1945 some 200 BW projects were in progress under the programme for Fiscal Year (FY) 1946. During the six months that followed, Special Projects Division activities were greatly curtailed. The US BW research and development facility at Vigo was placed on stand-by, and Horn Island and the Granite Peak installations fully decommissioned. The end of the War ushered in an era of review and reorganisation of US BW activities. The Chief of the CWS was requested to submit a peacetime research and development programme, and personnel involved in research and development activities submitted final reports related to their activities. As can be seen from Appendix I, security classifications were revised in order to permit the publication of research findings in the scientific literature. In addition to studies relating to phytopathology, the papers covered a broad spectrum of scientific inquiry that included: bacteriology, physiology, pathology, clinical medicine, preventive medicine, biochemistry, neurology veterinary medicine, mycology, botany, public health, industrial hygiene, instrumentation, chemical engineering, chemistry, and agriculture. By 1958 some 660 such papers had been published.[1]

In chapter 8 US investigations into problems associated with the large-scale production of plant pathogens as BW agents are set in the context of aspects of transatlantic collaboration on anti-crop BW. Chapter 9 then goes on to discuss the different means of dissemination for anti-crop BW agents under development in the US between 1947 and 1958.

However, the period under review in the section that follows corresponds with respective phases in the US offensive BW programme identified by Laughlin[2] and set out in Table 7.1:

Table 7.1 Phases of US offensive BW programme

Dates	*Activity*
1946–49	Research and planning years after the Second World War
1950–53	Expansion of the BW programme during the Korean War
1954–58	Cold War Years – reorganisation of weapons and defence programmes
1959–62	Limited war period – expanded research, development, testing and operational readiness
1963–68	Adaptation of the BW programme to counter insurgencies – the Vietnam war years
1969–73	Disarmament and phase down

The post-war period saw the Soviet Union emerge as a powerful ideological and military rival to the United States. Soviet ideological and military domination would extend across Eastern Europe, and the USSR would demonstrate its capability to produce nuclear weapons and thereby pose a credible military threat to the US and its allies in the North Atlantic Treaty Organisation (NATO).

The extent to which the Soviet Union had made progress in the field of BW would remain the subject of some speculation into the mid-1950s. Compiled in 1949, the CIA's *Quarterly Review of Biological Warfare Intelligence*[3] could only offer the basic assumption that in scientific and industrial terms the Soviet Union was capable, as the report states:

> of maintaining a program of research and development in BW comparable to that of the US... . It is believed that the USSR ought to be capable of offensive use of BW both in sabotage and in open warfare.

Joint Intelligence Committee reports on 'Soviet Capabilities for Employing Biological and Chemical Weapons'[4] produced in January of the same year set forth the locations of suspected Soviet installations with possible BW connections but concluded that due to 'extremely effective security' no definite evidence on Soviet BW research and development facilities was available.

British documentation[5] pertaining to Soviet BW capabilities during this period also revealed a dearth of reliable intelligence data about Russian developments in the scientific and technical fields, 'which might indicate the scope of future developments'. Further British documentation relating to Soviet BW capabilities also contained information of a speculative nature. The 1952 Report[6] by the UK Biological Warfare Sub-Committee could only offer the following conclusion:

> 'There is no firm evidence of the existence in the USSR of any BW project either for research, mass-production of BW agents, or the development of the necessary special weapons and equipment. There are however indications that a small group of scientists may be engaged on BW research under the control of the Soviet Army. The only broad conclusion possible is that the Russians are now capable of BW sabotage wither against man, livestock, or crops, and that they could, if such were their intention, have initiated the mass cultivation of bacteria in 1951 and achieve by 1952 at least the level of production attained by the U.S.A. in 1945.

Perceptions of Soviet BW capabilities would gradually increase in US political and military circles toward the end of the period presently under review to the extent that US CW and BW policy would be brought into question more and more.

Research and planning years after the Second World War (1946–49)

The FY 1947 BW project programme revealed a consolidated programme of research and development. According to one partially declassified report,[7] the post-war programme was organised into the following activities:

> 1. Screening of bacteria, viruses, fungi, protozoan, and helminth parasites for potential biological warfare agents.
> Studies of anthropods as biological agent vectors.
> Laboratory development of selected agents.
> Development of biological defensive measures against selected agents.
> Clinical studies of diseases encountered in biological warfare research.
> 2. Development of pilot plant equipment for production and processing of selected agents.
> Studies of the production and processing of selected agents.
> 3. Screening, development, production, and protection against plant inhibitors and defoliants.
> 4. Development of aerosol munitions, surface-contaminating munitions, guided missiles, small arms, and covert methods.
> Design of methods for filling biological warfare munitions.
> Surveillance studies of munitions.
> Tactical employment of munitions.

5. Decontamination materials and methods.
6. Protective clothing development.
7. Studies of biological warfare occupational hazards.
8. Nutrition of potential agents.

Cloud chamber studies.
Studies of the special physiology, pathology, and pharmacology of biological warfare diseases.
Study of chemical methods in biological warfare research.
Study of meteorological factors incident to biological warfare.

Close liaison between respective agencies involved in the field of BW was maintained during the course of this period with the offices of the Surgeon Generals of the Army and Navy, Ordnance, General and Special Staff Divisions, the US Public Health Service, the US Department of Agriculture, and the National Institute of Health all participating. Close collaboration was also maintained with agencies in the UK and Canada.

A notable characteristic of the US programme as it developed at Detrick related to the initiation of a cloud chamber project which began in 1943, and was fully operational by 1945. According to SIPRI:[8]

> This cloud chamber project, in addition to the less elaborate experimentation along similar lines in the UK and Canada, provided a mass of data establishing some of the mechanisms of airborne infection. Once these methods had been elucidated, it became possible to assess the feasibility of creating unnatural forms of airborne infection, namely the feasibility of effective dissemination of BW agents in aerosol form.

An Advisory Committee on BW Research and Development was established in 1945 to operate in an advisory capacity to the Chief of the CWS on matters relating to the progress of the biological warfare programme. Under further reorganisation in 1946 the Joint Research and Development Board (JRDB) assumed the responsibility of the Advisory Committee on BW. Progress reports and recommendations were passed back up the chain of command from various sub-panels via Executive Secretaries, and on up to the Chief of the CWS via the JRDB. With further reorganisation of the National Military Establishment which saw the formation of the US Department of Defense in 1949, the JRDB became the Research and Development Board (RDB) under the office of the Secretary of Defense with Vannevar Bush as Chairman. The RDB based its appraisal of the US BW programme primarily upon reports submitted by the Committee on BW. That Committee submitted technical estimates relating to anti-personnel, anti-animal, and anti-crop BW munitions. According to one partially declassified report:[9]

> the period between 1947 and 1952 was an era of boards, committees, Ad Hoc groups, panels, contractors, etc. investigating, evaluating, and advising on various phases of the BW program. (At one time during a period of a few months, 23 such groups were engaged in studies and surveys.)

With post-war Chemical Corps work in the fields of CW, BW and radiological warfare ongoing, it was not until July 1948 that a comparative evaluation was commissioned by the Joint Chiefs of Staff.[10] While the results of this review were not made available to this author, an interim report by the Chairman of the RDB revealed that formal review of CW, BW and radiological warfare was long overdue, and that lack of reliable data had hampered efforts to give serious consideration to these forms of warfare at the level of military planning and policy.

In the same year in the field of BW, Secretary of Defense, James V. Forrestal, had commissioned a Special Ad Hoc Panel on Sabotage to review US vulnerability to covert BW attack. The Panel submitted its Report on Special BW Operations in October.[11] The Report included the following observations regarding US vulnerability, that:

> 1. Biological agents would appear to be well adapted to subversive use.
> 2. The United States is peculiarly susceptible to attack by special BW operations.
> 3. The subversive use of biological agents by a potential enemy prior to a declaration of war presents a grave danger to the United States.
> 4. The biological warfare research and developments program [does not now] meet the requirements necessary to prepare defensive measures against special BW operations.

The report detailed a series of research and development objectives for defensive and offensive special BW operations with one defensive objective recommending a series of vulnerability tests with innocuous organisms, on crops, food, livestock, subway systems, ventilating systems and water supply systems. Such testing, according to the Report, would help 'to determine quantitatively the extent to which such subversive dissemination of pathogenic biological agents is possible'.[12] The Report concluded as follows:[13]

> It is recommended that the Secretary of Defense authorize the National Military Establishment to engage in research and development in the field of special BW operations and direct the Research and Development Board to assign responsibility for this program. Such directives are necessary to initiate research and development in this field.

These recommendations would form the basis of post-war collaboration between the Army and the intelligence services in the field of special BW operations.[14] According to Laughlin,[15] the Report's recommendations were:

> subsequently approved and became the genesis of open air vulnerability tests and covert R&D programs conducted by the Army, some of which were in support of the Central Intelligence Agency (CIA).

In May 1949, in accordance with the recommendations included in the report, a Special Operations Divisions was established at Camp Detrick.

Similar studies[16] on clandestine threats to the UK and the countries of the Commonwealth revealed that such attacks posed a significant threat to man, animals and crops.

The Haskins Committee

A predominantly civilian Ad Hoc Committee had been appointed by Secretary of Defense, Forrestal, to conduct a review of BW in March 1949. The purview of the Committee, however, was subsequently extended to include under its review chemical, biological, and radiological weapons – collectively referred to hereinafter as 'CEBAR'. The Committee, chaired by Chemical Corps officer Caryl P. Haskins, submitted its review entitled, 'Report of the Secretary of Defense's Ad Hoc Committee on Biological Warfare'[17] in July 1949. The report, presently only partially declassified (hereinafter referred to as the Haskins Report), numbered some 38 pages and was organised into the following sections: Letter of Transmittal to the Secretary of Defense; Major Findings; Recommendations for Action by the Secretary of Defense; Background; and, Scope of Biological Warfare. A section concerning Major Issues: Discussion and Specific Recommendations was organised into the following sub-sections: 1. Biological Warfare – Defensive versus Offensive; 2. Sabotage with CEBAR Weapons; 3. CEBAR Intelligence Requirements;

4. Military Applications; 5. Economic Consequences of CEBAR; 6. Human Epidemics; 7. Research and Development Policies; 8. Public Information Policies on CEBAR; 9. Defense Against CEBAR Warfare; 10. Political and Psychological Implications of Biological Warfare, and; 11. United States Position on CEBAR in International Affairs. The available sections and sub-sections of the report will discussed in more detail in the paragraphs that follow.

The Chairman of the Secretary of Defense Ad Hoc Committee on Biological Warfare outlined the Committee's objectives in a Letter of Transmittal[18] to the then Secretary of Defense, Louis Johnson, as follows:

> Your Committee has considered the offensive and defensive potentialities of biological weapons, clandestinely or overtly... . The Report deals with the present capabilities of CEBAR weapons and emphasizes their future potentialities, neither of which appear to be fully appreciated in military or civil defense planning.

In the section dealing with 'Major Findings,' the Haskins Committee included an appraisal of the potency and future potential of CEBAR warfare. Under the section entitled 'Recommendations for Action by the Secretary of Defense,' the Committee put forward a series of recommendations that, if adopted, would have propelled CEBAR from a position of relative obscurity in military and defense planning, to being given formal consideration at the highest levels of the National Military Establishment. In this connection the Committee put forward the following recommendations,[19] that:

> Adequate emphasis should be placed on CEBAR within military planning. This includes both strategic planning and continuing support by the National Military Establishment of research and development on military aspects of CEBAR. [The Secretary of Defense should] establish with this immediate office the position of a personal CEBAR Advisor.
>
> The function of the CEBAR Advisor should include the following:
> 1. Advise the Secretary of Defense on CEBAR matters.
> 2. Initiate and encourage a broad public information program on CEBAR.
> 3. Stimulate strategic thinking in this field within the Joint Chiefs of Staff and the three Services.
> 4. Assist the Secretary of Defense in arranging inter-agency co-operation in defense problems arising from the possible use of CEBAR.

5. Assist the Secretary of Defense in promoting a program of research on public health and other defensive measures against CEBAR.
6. Assist whenever possible in expanding the existing research program on military aspects of CEBAR.
7. Encourage and assist the appropriate intelligence agencies in the development of adequate CEBAR intelligence.
8. Represent the Secretary of Defense in inter-agency discussion relative to the US international position on CEBAR matters.

A recommendation to promote civil and military defensive preparations was also put forward by the committee.

The section entitled 'Background' included an explanation for considering chemical, biological and radiological warfare together, as follows:[20]

> The report deals primarily with biological warfare ... and secondarily, with chemical and radiological weapons when the common characteristics of the weapons indicate that common policies and applications are desirable. From the standpoint of biological warfare, such combined use is likely to increase effectiveness, and therefore, the Committee has emphasized that these weapons should be considered collectively when policies and strategic decisions are involved... .

The Haskins Committee found US offensive BW in a poor state of preparedness with no biological weapons available for immediate use and only prototype models of munitions at various stages of research and development. Haskins reported that the facilities and techniques existed to produce BW weapons with limited effectiveness against man, animals and plants, and that such capabilities could only be deployed after a period of six to nine months once a decision to go into full production had been made.

In the section entitled 'Scope'[21] it was concluded that CEBAR offered an opportunity to attack man:

> directly as a biological organism through the spread of infectious or contagious disease, through use of chemical or radioactive poisons, or indirectly by: A. diminishing his food and other living resources. B. disorganising his social fabric by damage to vital national and community facilities [and]. C. undermining his morale by exploiting CEBAR's 'anxiety potential.

The Committee emphasised what it referred to as the 'unique versatility' of CEBAR warfare against man with effects ranging from temporary incapacitation to prolonged illness and death, and CEBAR's 'high degree of strategic and tactical flexibility'. The Committee also reported that both animals and plants were particularly vulnerable to CEBAR warfare. According to the report:[22]

> Disastrous consequences would result from the destruction of the principal cereal crops of large segments of the populations of China or India. Similarly, in Central Europe, the destruction of the potato crop would have serious effects.

While the Special Ad Hoc Panel on Sabotage warned of US vulnerability to covert BW attack, the Haskins Committee expressed the view that Soviet developments in the field of CEBAR might, if coupled with success in the field of atomic weapons, disrupt the prevailing balance of power and precipitate conflict between East and West:[23]

> The Committee wishes to call attention to a consideration of particular importance having a major bearing on the international balance of power. Accurate appraisal of this matter will require much wider information, particularly of an intelligence nature, than has been available to the Committee. This is the bearing which the Soviet estimate of its own military capabilities vis-à-vis those of the rest of the world has on its courses of action. Such Soviet self-appraisal will affect both the emphasis given to the development of weapons of biological warfare within the U.S.S.R., and Soviet policy with respect to international armaments agreements. Thus the Soviets might well feel that a major technical advance in biological warfare in their own country could put them in a position of weapons parity in this field. On this basis, they might expect to press for the elimination of atomic weapons in return for the prohibition of biological weapons. Further, the development of effective CEBAR weapons, particularly if accompanied by success in the atomic field, might lead the Soviets to believe that they possessed over-all weapons superiority and that the time had come to precipitate openly 'the inevitable conflict between capitalists and communist societies.

Whilst the sub-sections concerning 'Major Issues: Discussion and Specific Recommendations' have not been made available to this author, the Haskins Committee confirmed a lack of guiding policy and found in favour of the implementation of a wide-ranging programme in the field of CEBAR warfare. Joint Chiefs of Staff (JCS) action on the Haskins recommendations, however, was delayed pending the completion of a follow-up investigation into CEBAR,[24] and according to Miller,[25] '[n]o positive action was taken on the findings and recommendations of this committee.

Expansion of the BW programme during the Korean War, 1950–53

Matters relating to the US CW and BW programme had been the subject of discussion in both political and military circles in the later half of the 1940s. On the one hand, respective House of Representatives and Senate appropriations committees heard Chemical Corps testimonies relating to budget requests. While on the other hand, under conditions of secrecy, deliberations relating to the role of chemical warfare in US military posture took place at the highest political levels. While Roosevelt had declared a clear no-first-use policy for US CW weapons during the course of the Second World War, no statement relating to US BW policy was forthcoming, and declared US policy with regard to the employment of BW and radiological warfare would remain unclear until the latter half of the 1950s. According to Wright:[26]

> Until 1956 the US Army Field Manual 27-10, 'The Rules of Land Warfare,' stated that gas warfare and bacteriological warfare are employed by the United States against enemy personnel only in retaliation for their use by the enemy.

In an interim policy review in 1950 Truman reaffirmed the US CW policy of retaliation only[27] but this affirmation was subsequently brought into question by a further Ad Hoc Committee reporting on CW, BW and radiological warfare.

The Stevenson Committee

Under the Chairmanship of a civilian Earl P. Stevenson,[28] a Committee was appointed to report on US CEBAR warfare policy and preparedness by Secretary of Defense, Louis Johnson, in December 1949. The Committee's 'course of action' was planned at a two-day initial meeting in January 1950 at Washington DC, where presentation where heard from: Rear Admiral A. C. Davis on current thoughts of the Joint Chiefs of Staff on CW, BW, and radiological warfare; Major General A. C. McAuliffe, who explained the responsibilities of the Chemical Corps in these three fields; Dr. Vannevar Bush, who spoke on the varied facets of the problems facing the Committee; Dr. W. Albert Noyes, Jr, who gave a resume of the other studies that had previously been undertaken in the area of interest to the Committee; and from the Honourable Marx Leva, Assistant Secretary of Defense, who outlined the necessity for the establishment of the Committee.[29] Over a period of some six months the Committee then met a further nine times. The Committee submitted its findings in its Report of the Secretary of Defense's Ad Hoc Committee on Chemical, Biological and Radiological Warfare, 30 June 1950 – five days after the outbreak of war in Korea. Preceded by its Letter of Transmittal to the Secretary of Defense, the Committee's 37 page Report was organised into the following sections: Forward; and Major Recommendations. A section entitled 'Findings and Conclusions' was organised into the following sub-sections: Misconceptions; National Policy; CW; BW; Radiological Warfare; Public Information; and Psychological Aspects. The latter sub-sections were followed by the inclusion of five appendices: A. Synopsis of the Committee's Activities; B. List of Witnesses; C. Individuals Interviewed by Sub-Committees or by Individual Members of the Committee; D. Documents Prepared for the Committee; and E. Other Documents Considered by the Committee. The relevant sections of the report will be discussed in the paragraphs that follow.

In the 'Letter of Transmittal' to the Secretary of Defense, Stevenson outlined what the Committee had identified as, 'significant gaps in thinking and programming'[30] with regard to CEBAR warfare. The committee ruled out the likelihood of effective international control of CEBAR warfare, and stated that no moral distinction could be usefully made between CW, BW and radiological warfare. It emphasised the low level of US toxic CW preparedness, the utility of such weapons in the face of advancing enemy ground forces, and the threat posed to the US and its allies by substantial production and stockpiling of toxic chemical agents by the Soviet Union. The most significant recommendation of the Committee was made in relation to the policy of no-first-use. '[U]se in retaliation only', the Committee held:[31]

> has resulted in the assignment of low priorities to the research, development, and production of chemical weapons. The security of

> the United States demands that the policy of 'use in retaliation only' be abandoned.

The report went on to emphasise the low level of priority that had been assigned to offensive and defensive aspects of BW and radiological warfare. The Letter of Transmittal offered the following conclusion:[32]

> The United States is not prepared for biological warfare and, for all practical purposes, is not prepared for chemical warfare. This state of unpreparedness is the result of the indecision that, during the post-war years, has permeated the area of the Committee's investigation. We believe the recommendations that we make would, if accepted, serve to break the deadlock and produce action. To carry out these recommendations will require an outlay of additional funds, but the relatively small cost appears to be a sound investment.

Under Appendix D, 'Documents Prepared for the Committee', Col. William M. Creasy, Cml C, Chief, Research & Engineering Division, had submitted a presentation to the Secretary of Defense's Ad Hoc Committee on CEBAR. Creasy's presentations, submitted 24 February 1950, as outlined by the Stevenson Report, included a, 'general indoctrination of the entire field of biological warfare', and was organised into the following sections: 1. Introduction and Historical Background; 2. BW agents – anti-personnel, anti-animal and anti-plant; 3. Production of BW agents; 4. Dissemination of Biological Agents; 5. Defensive Aspects; 6. Offensive Employment of Biological Warfare (including tactical, strategic and sabotage operations); 7. The Logistic and Operational Implications of a BW Attack with Anti-Personnel Agents; 8. Research and Development Activities in Biological Warfare; and 9. Co-operation with Canada and the United Kingdom. Creasy's presentation offers an interesting overview of developments relating to BW against man, animals and crops and the general state of US BW preparedness in the period immediately preceding its submission to the Secretary of Defense's Ad Hoc Committee on CEBAR. Creasy's presentation is discussed in more detail in the paragraphs that follow.

In 1950 the principal US BW research and development facility was located in the Biological Department[33] at Camp Detrick, Maryland. The Biological Department functioned as an operating agency of the Research and Engineering Division of the Chemical Corps[34] and was staffed by approximately 656 civilian (professional and non-professional) personnel, and 96 Air Force, Army, and Navy personnel.

During FY 1950, the US BW research and development programme consisted of 87 projects conducted by the Biological Department costing of $5,600,000.

Anti-personnel BW agents reported to be under consideration included *Brucella suis,* Psittacosis Virus, *Bacterium tularense* [sic], the latter presumably a reference to *Francisella tularensis,* (the causal agent of rabbit fever), and *Bacillus anthracis.*[35] Although Creasy reported that no anti-animal agents were in the 'developmental stage' both Rinderpest virus and the virus causing foot-and-mouth disease were reported to be two of the most effective anti-animal BW agents. With regard to anti-crop agents under investigation during this period, Creasy discussed developments relating to two types of agents: 1. Chemical Plant Growth Regulators, such as 2,4-D; and 2. Plant Pathogens, represented by bacteria, parasitic fungi, and viruses. In regard to biological anti-crop agents, according to Creasy:[36]

> All three groups are being considered as potential agents. Of the three groups, the parasitic fungi are the most important and are, at present, receiving the most attention. They are represented in part by the smuts and rusts of the grain fields. Such diseases are specific to the cereal crops such as rice, wheat, oats, and corn. They account for a loss of greater than 100 million bushels of grain annually in the United States through natural infection. Another example of this group which does not infect the cereal crops is the late blight fungus of potatoes.

It was reported that, if required, full-scale production of anti-personnel agents such as anthrax and botulinum toxin could be achieved after a period of one year with the Vigo Plant at Terre Haute, Indiana, capable of producing anthrax in quantities sufficient to fill 500,000 E48R2 bombs, or quantities of botulinum toxin sufficient to fill 250,000 bombs per month, using the 'standard four day cycle for growth.'[37] According to Creasy, the pilot plant at Camp Detrick, although used primarily at this time for research, could be activated to produce sufficient quantities of *Brucella suis* and the causal agent of Tularemia for the filling of 14,000 E48R2 bombs per month. The virus production facilities at Camp Detrick was also capable of producing significant quantities of psittacosis and other viruses.[38] It was reported that, 'probably the most advanced phase of BW production is in the field of chemical plant growth regulators'.[39] According to Creasy, '[s]tockpiling of 5 million pounds of 2,4-D has been recommended, but at the present time only a few thousand pounds of the agent are stored at Deseret, Utah'.[40]

In the event of an emergency it was estimated that anti-personnel agents could be produced in sufficient quantities to mount an attack that would, 'place an effective dosage over 90 square miles of enemy territory. This area is comparable to that of the city of Washington, D. C. Thereafter, we would be able to mount attacks of like magnitude every four days'.[41] Production facilities were capable of producing 8 million lb of the chemical plant growth regulator 2,4-D per quarter. With regard to the production of anti-crop plant pathogens it was reported that:[42]

> Production of plant pathogens is now considered feasible. It has been determined that one ton of spores may be harvested from 80 acres of infected cereal growth. If the need existed, it would be possible to secure sufficient quantities of plant pathogens to carry out retaliatory strikes in approximately 6 months.

Research into the dissemination of anti-personnel agents was reported to have focused on experimentation into the behaviour of aerosol clouds, establishing a criteria for lung-retention sized particles (developers worked between the limits of 1 and 5 microns[43]), and experimentation relating to the introduction and retention of such particles in the respiratory system. According to Creasy:

> BW munitions must produce aerosols within a desired size range, of viable organisms which are at a concentration sufficient to produce the desired dosage during exposure time of the population of the aerosol cloud.

Although the Creasy presentation noted that no ' ... standard munitions for BW agents are available at the present time', the E48R2 4lb Particulate Bomb was reported to be nearing the final stage of development (104 E48R2 4lb bombs were clustered in an E38-type 500-lb cluster adapter to make the E96 cluster).

When dropped from an altitude of 35,000 feet, this munition could generate an elliptical aerosol cloud of anti-personnel or anti-animal agent with major axes of 3000 feet and 800 feet.

Other munitions were reported to be at a more basic stage of research and development and were listed as follows: 20-mm Particulate Projectile; $\frac{1}{2}$-lb Particulate Bomb; Continuous Aerosol Generator; Flettnor Rotor; and, a BW Warhead for Guided Missiles.

Experimentation had taken place into achieving surface contamination with vapour clouds of anti-plant chemical agent, and experimentation into aerosols of chemical agent had indicated the, 'practicability of using a continuously generated aerosol from a simple munition comparable to a coloured smoke grenade or screening smoke bomb'.[44] Research and development into achieving a means of dissemination for anti-crop plant pathogens was described as follows:[45]

> A static trial has recently been made with a self-propagating epiphytotic agent dispersed from a 20-mm particulate projectile in a field of growing grain to produce surface contamination. Results indicate that the spores, so disseminated, have been deposited on the grain and that spore colonies have grown and have rapidly spread through the remainder of the field.

The 20-mm Particulate Projectile was described thus:

> The 20-mm machine gun shell, modified for dissemination of BW agents, has a sensitive nose fuze and a powder charge producing base expulsion of the agent filling on impact. It is designed to produce aerosols of anti-personnel and self-propagating anti-crop agents. ... The agents causing self-spreading diseases of plants require a munition which will create a large number of small foci of infection in a regular pattern over a large area. The amount of agent required to initiate each small focus of infection is very small and can be contained in a 20-mm projectile.

A description of the characteristics of BW Munitions Under Development as described in the Creasey Report related to the following munitions: the E482R anti-personnel/anti-animal bomb; the 20mm anti-personnel, anti-animal, and anti-crop projectile; the 1/2 lb anti-personnel/anti-animal bomb; a continuous aerosol generator; and a Fletnor Rotor for the uniform dissemination of BW agents over a wide area.

Under Section VI, 'Offensive Employment of Biological Warfare', Creasy assigned a strategic role for offensive use of anti-crop BW weapons, which were 'strategic by the very nature of the targets'. In accordance with the US Air Force definition, strategic operations included, 'the destruction of the enemy's military, industrial, economic and political system and undermining morale of the enemy people to the point where their capacity for resistance is weakened fatally'. A capability to wage anti-crop BW was reported by Creasy to be attainable but dependent upon the level of research and development assigned to this field:[46]

> Our potential enemy relies heavily upon cereal and other crops for a major part of his food supply. BW attacks against his crops are a very attractive possibility and should be of great value in destroying his ability to wage war. Further development will be required before we shall be able to mount such attacks on a large scale. However, we do know enough now to think the goal of effective operations against enemy crops can be attained.

Having received the approval[47] of Secretary of Defense, Marshall, all but two of the eight recommendations put forward by the Stevenson Committee, resulted in the implementation of measures by the Joint Chiefs of Staff. According to a Memorandum[48] for the Chief of Staff, US Army, Chief of Naval Operations, and the Chief of Staff, US Air Force, dated 12 February 1951, Recommendation II approved that, 'necessary steps be taken to make the United States capable of effectively employing toxic chemical agents at the outset of a war'. Recommendation III approved that, 'construction be undertaken as soon as possible, followed by operation of a plant to produce militarily significant quantities of G-agents (nerve gases), and that munitions and means of delivery for these agents be brought to a commensurate state of readiness'. Recommendation IV approved that, 'projected engineering studies and designs of facilities for the production of biological warfare agents be completed as soon as possible'. Recommendation V approved that, 'field tests of biological warfare agents and munitions be carried out as soon as possible on a scale sufficient to determine the military worth of agent-munition combinations, their offensive uses and the means of defence against them, and to secure definitive information on other problems inherent in biological warfare'. Recommendation VI approved that, 'research programs on the defensive aspects of biological warfare be materially expanded'. Action on Recommendation VII on Radiological Warfare Evaluations was deferred. Recommendation VIII approved that, 'a co-ordinated program be established to guide releases of information on chemical, biological and radiological warfare, and that this program be reviewed and revised periodically'.

The Stevenson Committee recommendation (Recommendation I), that US no-first-use policy be abandoned, was rejected by the Secretary of Defense, but the period coinciding with the outbreak of war in

Korea saw an expansion of the US BW programme and increased efforts to achieve a retaliatory BW capability.

Air Force participation in the BW programme

Chemical Corps budget requests jumped from $25 million in 1949, and $32 million in 1950, to a total of £127 million under the Mutual Defense Assistance Program for FY 1951. Construction of a BW production facility at Pine Bluff Arsenal, Arkansas, received authorisation and limited field testing was resumed at Dugway Proving Ground, Utah, bringing a five-year period of inactivity to an end. Research and development facilities at Camp Detrick were also expanded during the course of period. With the expansion of the programme, reviews of US chemical and biological warfare preparedness continued and a directive to achieve a higher degree of chemical and biological warfare readiness was issued by the Secretary of Defense in December 1951.[49] In 1953 the respective chemical and biological elements of Chemical Corps activities were reorganised into separate entities and an Assistant Chief Chemical Warfare Officer for biological warfare was appointed. A further review of chemical and BW readiness was undertaken in the Summer of 1953. According to Laughlin:[50]

> In June 1953, a month before the Korean War ended, The Secretary of Defense expressed concern over the state of CBR readiness and stated that each Service, singly or in combination, should be prepared to employ CBR weapons when directed. After a review of the Services' capabilities, it was concluded that BW capabilities were, indeed, limited for a variety of reasons but primarily by knowledge gaps in the biological sciences.

The US BW programme was a joint undertaking involving Army, civilian, Navy, and Army Air Force (AAF) personnel. Primary responsibility for the US BW programme, which included the procurement, storage and issue of biological munitions, had been assigned to the Department of the Army in 1948.[51] According to Miller:[52]

> Subsequently, each service proceeded unilaterally in its planning. There was no specific assignment of joint responsibility in the fields of BW planning, doctrines, procedures, development of equipment, training, or budgeting.

The annual budget was requested and defended by the Chemical Corps. The Navy also provided financial support for the BW programme, and approximately one quarter of the technical staff required at Camp Detrick and other test installations, with personnel which included 8 officers and 70 enlisted men permanently stationed at Camp Detrick, recruited principally from the Navy Bureau of Medicine and Surgery, and the Bureau of Ordnance.[53] In February 1951, the Joint Chiefs of Staff issued a memorandum which clarified the respective roles of the Army, Navy, and Air Force with respect to the implementation of the BW programme.[54]

The United States Air Force had been identified as the major using agency for the aerial delivery of BW agents. Of the three types of weapons grouped under the acronym of CEBAR the Director of Research and Development, USAF, Major General D.L. Putt, had declared in September 1950 that: 'the US Air Force (USAF) is most interested in biological ... agents'.[55] In relation to the potential of anti-crop BW weapons the recognised capability of such weapons included by this time, 'reduction of yield of crops to disaster proportions'.[56]

Prior to the establishment of the Air Force on an autonomous basis, however, AAF participation in the US BW programme was one of co-ordination. Closer liaison developed between the Chemical Corps and Air Force with the Air Force assuming responsibility for stating technical requirements for BW munitions. Miller outlines the level of Air Force participation in the BW programme in 1949, as follows:[57]

> By March 1949, Air Force support for the BW program consisted of the air weather detachment assigned to the Chemical Corps for research in micrometeorology and five liaison officers assigned to Chemical Corps. The Air Force also was furnishing aircraft, flying equipment, flight clothing, and air base facilities for testing activities.

Pending the submission of the Weapons Systems Evaluation Group (WSEG) report[58] on CEBAR warfare, which was not expected until 1954, the Air Force commissioned an interim evaluation of BW in September 1950. The study, undertaken by the Rand Corporation, evaluated the requirements for an Air Force minimum offensive BW capability. Air Force preparedness in BW, however, was found wanting. According to Miller:[59]

> The USAF had no production plants for agents, no stockpile, and no experience in employing BW agents. The situation provided no current capability whatsoever with Biological Warfare munitions.

As identified by Miller, a number of factors combined to contribute to the state of Air Force preparedness described above. Military spending on BW research and development was insignificant in comparison to other major weapons development programmes, and in FY 1951 the budget figure was $6,000,000, or approximately 2 per cent of the total military budget for research and development. It was not until fiscal year 1950, however, that the USAF began to budget for its own research and development programme. Even then, according to Miller:[60]

> Money expended to meet Air Force requirements for BW weapons from 1947 to July 1951 totalled only $2,971,692.

For FYs 1950 and 1951 Camp Detrick had received $11,712,094 for research and development from other sources while for the same period the Air Force contribution was only $2,154,000. While the slow transfer of funds from Air Force to developing agency acted as a brake on Air Force projects assigned to Camp Detrick, as Miller has observed, 'USAF support was only a small part of the entire program'.[61]

BW research and development was conducted under conditions of secrecy rivalled only by the Manhattan Project on nuclear weapons and in spite of the introduction of more liberal security procedures by the Joint Chiefs of Staff in 1948,[62] the secrecy surrounding the venture hampered progress in research and development. In the sphere of military indoctrination, trained Air Force BW and CW personnel numbered only 10 in January 1951. In a report submitted in July 1948 to Major General E. E. Partridge, Director of Training and Requirements, USAF, lack of recognition of the potential of BW among the upper echelons of the USAF had been cited as the reason for low levels of Air Force participation in the US BW programme. In a follow up study submitted to the USAF Chief of Staff in March 1949, an Air University Staff Study on BW concluded that, 'BW was then being neglected by the National Military Establishment, and particularly by the Air Force.' On the one hand, Air Force preparedness in BW was hampered by lack of strategic guidance from the Joint Chiefs of Staff. While on the other hand, the policy of 'use in retaliation only', which prioritised military preparations for conventional warfare with high-explosive munitions, accorded a low priority to CEBAR warfare. As stated by Miller,[63] Air Force participation in the BW programme up to 1950 was characterised by:

> a curious mixture of an awareness of the importance of the program, with an apparent lack of positive, aggressive action. Certainly, the

> program had to have the complete support of the Joint Chiefs of Staff if anything was to be done about it. It seemed, however, that high level decisions made during the transition of the Air Force into a separate department did little to promote the program.

However, with the implementation of Joint Chiefs of Staff recommendations regarding BW preparations, the functions and responsibilities of the Assistant Deputy Chief of Staff, Operations, were expanded, according to Miller[64] to, 'include those incident to the earliest attainment of an Air force-wide combat capability in BW–CW on a high priority basis'.[65]

The significance of the latter recommendation was considerable in that it consolidated the relationship between the USAF and the US BW programme.

The Research and Development Board had established a criteria for the standardisation of BW munitions in August 1949, and the development of anti-personnel agents was accorded top priority. USAF emphasis on the development of BW weapons also resulted at this time from the recommendations made by the Stevenson Committee.[66] A strategic category of 1 had been assigned to the BW programme in February 1951 by the Joint Chiefs of Staff with the USAF charged with responsibility for developing a worldwide combat capability.[67]

The following seven anti-personnel and anti-crop BW agents had been judged 'feasible for use' by October 1950: *Bacillus anthracis*; *Brucella species*; *Brucella tularense*; *Bacillus pestis*; *Botulinus toxin*; cereal rust spores (plant pathogens); and chemical plant growth regulators.

With regard to anti-crop BW research and development, bacteria, viruses and parasitic fungi, continued to be screened for their disease producing potential against a variety of crops with emphasis placed upon parasitic fungi as BW agents against cereal crops. In this connection the earliest anti-crop BW capability was not envisaged to be available until March 1951 whereupon, according to Miller, 'it was estimated that 50 per cent of Russia's winter wheat crop could be destroyed on a single mission of nine B-29 equivalents'.[68] However, progress in the development of anti-personnel and anti-crop warfare proved to be slow and the development of doctrine, operational plans, logistics and procedures lagged behind progress that had been made in the development of munitions. Further guidance in respect of the progress of the USAF BW programme subsequent to the Stevenson Committee recommendations came in the first official statement of the establishment of a USAF biological warfare effort known as the Twining directive.[69] The Twining directive, which was issued on 15 January 1952, established a time-phased programme for the establishment of an early USAF BW capability. The directive specified that by December 1953 a one-wing Strategic Air Command (SAC) capability for the dissemination of anti-personnel, anti-crop and anti-animal agents from medium bombers would be expanded to three wings of medium bombers, and that by December of 1954 all Strategic Air Command Units would be BW capable.[70]

By 1953 a limited anti-personnel and anti-crop BW capability had been achieved with the standardisation of the M33 anti-personnel bomb filled with *Brucella suis*, and the M115 anti-crop BW bomb filled with the causal agent of wheat and rye rust. While the emergence of the above weapons met the short-term objective set out in the Twining Directive, both munitions were deemed to be of questionable military value due to their small area coverage. Additionally, difficulties associated with the logistical and operational use of such weapons placed undue demands on resources, personnel and training. The short-term objectives of the Twining Directive were abandoned in October 1953 and on the authority of the Secretary of Defence the USAF BW programme was re-orientated committing the Air Force to longer-term research and development objectives and the development of superior weapons.[71] Development of anti-personnel BW munitions switched from the M33, with its incapacitating payload, to the lethal E61 biological bomb filled with anthrax but in spite of the perceived superiority of this new weapon the survival of the USAF BW programme would remain in doubt until 1955 when the Joint Chiefs of Staff would be in possession of the long-awaited Weapons Systems Evaluation Group (WSEG) survey. Prior to 1955 the requirement to obtain a retaliatory capability went some way in ensuring the survival of the programme, however, in spite of the comparatively low level of investment in such weapons important progress in the field of anti-crop BW munitions provided a much-needed impetus in favour of the continuation of the programme.

Table 7.1 compares Developmental Funds on anti-personnel and anti-crop warfare munitions between Fiscal Years 1954 and 1956.

Cold War years – reorganisation of weapons and defence programmes, 1954–58

Re-orientation of the USAF BW programme resulted in emphasis being placed upon further development of the E61 bomb with its lethal anti-

Table 7.2 Projects in technical developments area in the US, March 1954[72]

Type	*Prior $ million*	*FY 1954 $ million*	*FY 1955 $ million*	*FY 1956 $ million*
Research and Development in BW munitions	5,722	1,350	775	1,000
Anti-personnel BW munitions	2,653	2,100	1,614	1,110
Anti-Crop BW munitions	612	350	264	200

personnel payload and on the development of weapons for the delivery of anti-crop BW munitions. Anti-personnel agents were to be developed for release from high-speed aircraft and anti-crop agents from high-speed aircraft and from a balloon delivery system. During re-orientation of the programme the services were required to maintain their existing capability in BW. In order to correct the deficiencies associated with the previous phase of the programme, training of key USAF BW personnel was intended to be integral to its re-orientation but in reality this aspect of the programme fell short of expectations.

Subsequent to the implementation of the Stevenson Recommendations the debate over the policy of retaliation only had continued. Both the USAF in 1952, and later on in 1954, the Army had submitted recommendations to the Joint Chiefs of Staff that the policy of retaliation only be revoked. Such a recommendation, however, brought with it a number of risks. One the one hand, the revoking of the no-first-use policy by the US might destroy any legal or moral restrictions that had hitherto prevented the Soviet Union from contemplating the use of such weapons. Additionally, a change in policy might imply the intention to use such weapons when many needed to be convinced that the USAF had a credible BW capability that would justify such a change. Whilst on the other hand as Miller[73] points out:

> removing this restriction would restore to the United States the initiative in the use of biological weapons – a prerogative which it had surrendered under the existing national policy.

By 1954 the USAF had revised its thinking regarding the change in policy and support for the policy of retaliation only became the official USAF position. By May of that year, however, a further major review of the retaliation only policy had been precipitated by the Chief of Staff of the Army.[74] High level national security advisors were to review the national policy on no-first-use in the light of Soviet pronouncements regarding the perceived use of chemical and biological weapons in future wars. Such pronouncements were first made in a speech given by Marshal Zhukov during the 20th CPSU Congress in 1956, and were repeated again three days later by the Commander-in-Chief of the Soviet Navy. According to Laughlin[75] US national policy on retaliation only was subsequently realigned thus:

> In 1956, a revised BW/CW policy was formulated to the effect that the US would be prepared to use BW of CW in a general war to enhance military effectiveness. The decision to use BW of CW would be reserved for the President.

Within a period of two years from the re-formulation of US policy on the first-use of CW and BW, however, US Army and Air Force collaboration on the development of anti-crop warfare weapons would be phased out due to decreasing interest by the Air Force.[76]

Limited war period – expanding research , development, testing and operational readiness, 1959–62

During the latter half of 1959 the Secretary of Defense sanctioned a five-fold expansion of the CW and BW programme and the Chief Chemical Officer was requested to prepare a five-year programme that included the revival of the anti-crop warfare programme. The Summer of the 1960 saw the revalidation of the revised policy of first-use. The development of CW and BW weapons during the course of this period saw emphasis placed upon their use in future limited warfare scenarios with the emphasis relating to the effects of such weapons shifting from lethality to incapacitation. Further reviews regarding the role of chemical and biological weapons in US policy ensued and after Joint Chiefs of Staff consultation with the Secretary of Defense a plan (Project 112) outlining the precise tasks, target dates and assigned action was formulated in September 1961. Implementation of the plan resulted in large increases in the US Army's Chemical Corps BW programme activity through to 1963.

Adaptation of the BW programme to counter insurgencies – the Vietnam War years, 1963–68

The onset of the Vietnam War saw developments in anti-personnel and anti-crop BW limited to maintaining previously stated production requirements for such weapons. However, during the course of this period to 1969 anti-crop agent production was accelerated under a contract that had been established in 1963 which centred around methods that permitted cultivation on a large scale. The programme nevertheless emphasised the production of anti-crop chemical agents, which were introduced in 1963 and used in large quantities in Vietnam until 1970.

With regard to the use of chemical defoliants in Vietnam, according to one source:[77]

> It was becoming clear ... that a new dimension had been added to warfare with its specific attendant risks of damaging the economy of the country, of upsetting ecological balances and of starving enemy troops, enemy sympathizers or just civilians in general.

Although the Vietnam War did not see the deployment and use of anti-crop BW, US research suggests that biological organisms could conceivably be used, under the right circumstances, as a means of warfare with effects of an order of magnitude greater than that of their chemical counterparts.

Disarmament and phase down, 1969–73

By 1969 President Nixon had renounced the use of lethal BW agents and weapons and all other methods of BW. Demilitarisation plans were then formulated regarding the destruction of the anti-crop BW stockpiles that were held at Fort Detrick, Rocky Mountain Arsenal (and prior to this at Edgewood Arsenal), and at Beale Air Force Base. The US anti-crop biological warfare stockpile of wheat stem rust spores and rye stem rust spores that had been produced and stockpiled at Edgewood Arsenal between 1951 and 1957, and the wheat stem rust spores that had been produced and stockpiled between 1962 and 1969 were destroyed by February 1973.[78] Stockpiles of the causal agent of rice blast disease that had been produced between 1965 and 1966 and stored at Fort Detrick were destroyed between January and May, 1972.[79]

Chapters 8 and 9 investigate: Anglo-American collaboration on matters relating to anti-crop warfare, the pathogens and their respective methods of production; and weapons development and testing.

8
Some Aspects of UK and US Anti-Crop Warfare Collaboration between 1943 and 1958

Transatlantic co-operation in CW and BW has a long and complex history. UK and US collaboration on research into CW, for example, can be traced back to the Great War of 1914–18, when, according to Carter and Pearson,[1] 'Anglo-American liaison underpinned the US Army Chemical Warfare Service created in 1918.'

Where collaboration between the UK and Canada on research into BW and on BW defence began during the late inter-war period, Anglo-American collaboration on research into BW began in December 1941, when the US entered the war.

Literature addressing chemical and BW collaboration is increasingly entering the public domain with one study[2] published in 1989, addressing transatlantic collaboration between 1937 and 1947 based primarily upon materials from Canadian archives. Other studies[3] published in the early 1970s, contain useful references to collaborative activity between the UK and the US, and a further study[4] published in 1996 presents an authoritative account of tripartite north Atlantic collaboration on chemical and biological warfare between 1916 and 1995.

The latter study contains a fascinating overview of the historical underpinnings of the often informal and *ad hoc* apparatus of chemical and biological tripartite collaboration prior to 1947. After the formalisation of this arrangement in 1947, it recounts the dates and venues of the tripartite meetings and the subsequent changes in British CW and BW policy during the course of the tripartite process.

Many of the transcripts relating to the technical discussion that took place during the course of the tripartite meetings, however, remain classified in the interests of national security, but much of the documentation relating to bi-lateral exchanges of technical information between constituent branches of the UK and US armed forces responsible for research and development into anti-crop CW and BW has begun to emerge into the public domain. The following account has been compiled from an incomplete set of publicly available official memoranda, minutes, and reports, relating to aspects of UK and US BW activities and therefore reflects the nature of the evidence available at the time of writing.[5]

The following review therefore relates to aspects of Anglo-American anti-crop warfare collaboration between 1943 and 1958. Although collaboration in anti-crop warfare extended to the sharing of information on both offensive and defencive aspects of this form of warfare, it is important to note that, for simplicity, the following description will concentrate only on offensive aspects of the respective anti-crop warfare programmes under review.

Historical underpinnings of transatlantic anti-crop warfare collaboration

Tripartite north Atlantic collaboration was an important feature in developments relating to the emergence of BW weapons in the US. In the period to 1947, informal co-operation between the US, Canada, and the UK emerged first with respect to developments relating to chemical warfare. Collaboration was subsequently expanded to include developments in BW concerning defence and retaliatory capabilities.

Collaboration between the US and the UK got underway with senior personnel from the UK visiting the War Research Service in the US and officials involved in BW research in Canada. The British BW programme, which had been established under the leadership of Dr Paul Fildes was started at the Biology Department, Porton Down, England, in October 1940.[6]

The legal position with regard to offensive BW research and development in the UK was clear. Although the UK had ratified the 1925 Geneva Protocol for the Prohibition of the Use in War of Asphyxiating, Poisonous or Other Gases, and of Bacteriological Methods of Warfare in 1930,[7] its pursuit of the acquisition of offensive BW weapons was not in breach of the Protocol which did not prohibit the development or production of such weapons. As Carter and Pearson note: '[T]he capability was seen as both a deterrent and an essential option in reacting to a BW attack.'[8]

Under Fildes' supervision two significant BW developments had emerged by 1942. First Fildes and his associates had established the required inhalation dose of anthrax spores to achieve infection and death in animals under laboratory conditions. And second, Fildes *et al.* had established a means of producing aerosolised[9] bacterial clouds of lung-retention sized particles. Safety considerations resulted in the requirement for testing to be subsequently moved out of the laboratory in 1942.

The testing with a BW agent and a means of dissemination was instituted under the supervision of Dr D.W.W. Henderson, Fildes' deputy, and Porton Scientist Dr D. D. Woods, at 'X-Base', Gruinard Island on the Ross-shire coast would remain the subject of controversy for some decades. In the summer of 1942 trails using modified 30 lb high explosive(HE)/Chemical bombs each containing approximately three litres of anthrax spores demonstrated the feasibility of infecting sheep with aerosols containing anthrax spores at a distance of up to 100 yards. According to Carter[10]

> [a] second trial on 24 July showed that the bomb was no less effective at 250 yards range. These trials demonstrated that the aerosol cloud was extremely potent, producing a lethal disease over nearly all of its front at 250 yards downwind. Data from sampling devices suggested that death could occur at 400 yards or more.

Trials were subsequently conducted with aircraft demonstrating varying degrees of success with the 30 lb bomb, and the programme switched to the use of specially adapted cluster munitions in 1943.

By 1943 conjoint Canadian/American BW research was underway at Horn Island, Mississippi Sound, and Canadian work was underway at Grosse-Ile in the St Lawrence River, and at the Suffield testing station, Alberta. The requirement for the harmonisation of an Anglo/American/Canadian retaliatory anti-personnel BW capability resulted in the development of a 500 lb cluster bomb known as the 'N' bomb containing over one hundred, 4 lb bomblets, which according to Carter and Pearson, was 'capable of producing a lethal aerosol of spores over about 100 acres from a small impact area'.[11] Large-scale production of spores was switched to production facilities in the US and further trials relating to the operational effectiveness of the 500 lb cluster munition were conducted in Canada with the aid of bacterial simulant.

In the absence of more rapid progress with regard to the conjoint 'N' bomb project, the British decided upon a alternative means of retaliation which was intended to undermine the agricultural sector of the German economy associated with the production of beef and dairy produce. Laboratory testing and field-trails had established the vulnerability of grazing cattle and anti-livestock cattle cakes were developed with the intention of causing anthrax infection. This development was of particular significance in the field of BW since it represented as Carter has pointed out, 'possibly the first acknowledged biological weapon'.[12] The British retaliatory capability was fulfilled by April 1943 with some 5,000,000 such cakes stockpiled at Porton, England.

In time representatives from both the US Army and the US Navy participated under the supervision of Fildes in the UK programme at Porton with one official returning from Porton to act as senior aid to Merck and another having participating in the BW field trials at Gruinard Island. The US programme is thought to have gained considerably from UK research on BW and information began to be exchanged between respective parties. According to Carter and Pearson:[13]

> One of the notable facets of British BW collaboration was Fildes' dispatch of a BDP (Biology Department, Porton) 'Green Book' on BW to Dr Ira L. Baldwin, the scientific director of Camp Detrick, and to Canada, in May 1943. The BDP 'Green Book' was an illustrated compendium of all Porton research on BW including the 1942 trails on Gruinard Island. Its provision to America and Canada set the seal on collaborative activity, and provided the emerging BW laboratories at Camp Detrick and in Canada with the totality of British experience and knowledge.

US and Canadian collaboration began with senior representatives from the Canadian BW programme known as the M-1000 Committee, liaising with WBC Committee members in 1941. The Canadian programme had begun with War Cabinet approval in 1940 and research was initiated at the Banting Institute, the Connaught Laboratories, and the University of Toronto.

Informal collaboration between the US and the UK on matters relating to CW had set an important precedent with regard to subsequent collaboration on BW. Reciprocal visits between US representatives of the Chemical Warfare Service at Edgewood Arsenal, and the British Chemical Warfare Committee had taken place as early as 1924 and such reciprocity appears to have been maintained throughout the 1920s and the 1930s. On CW issues American and British committees, such as the sub-committee of the British Inter-Services Committee on Chemical Warfare (ISCCW), had become fully integrated US/UK technical/administrative bodies – entities that remained distinct from those with responsibility for strategic planning and the formulation of policy. Britain had representatives stationed in the US, and the US stationed representatives in the UK in order to facilitate collaboration and the exchange of information. During the course of this early period US staff had also been stationed at the Canadian experimental station at Suffield, Alberta, Canada, and in Australia and India. Additionally representatives from the UK and Canada had been stationed at Dugway Proving Ground, Utah; and in Florida, and at the San Jose Proving Ground in Panama.

Regarding collaboration on BW, a similar pattern of exchange of personnel and information was facilitated through British representative Lord Stamp, who was accredited to the US Biological Warfare Committee (USBWC) and stationed in North America in 1943, and his opposite number, Lt Col. James H. Defandorf who became US BW liaison officer in London for the CWS. These apparent *ad hoc* arrangements were subsequently formalised under a process of tripartite collaboration which began in 1947. From the outset, however, the information-sharing arrangements established under the formalisation of collaboration did not appear to work well in respect of the sharing of information on developments in anti-crop warfare.

Although formal lines of communication had been clearly established through Lord Stamp in the US, and his opposite number in London, correspondence relating to an apparent break-down of communication on anti-crop warfare information was exchanged between British officials at the highest levels, and a US official operating in an advisory capacity to the US President.

Such correspondence is particularly significant in that, dated 26 June 1944, it contained a recommendation as to how the break-down of trans-Atlantic communication on matters relating to anti-crop warfare might be overcome. The correspondence proposed that such information be better channelled via the relevant US and UK agencies through the Inter-Service Committee on Chemical Warfare in Washington. In so doing, it recommended, such arrangements would facilitate, 'an opportunity for much better interchange than is operative at the present time'.[14] The correspondence went on:[15]

> if the matter becomes handled through the organisation here in Washington for the Combined Chiefs of Staff this is likely to be the most effective channel, for the parts of the subject which are not already interchanged, or of such nature as not to be readily interchanged.

War-time research, conducted by workers at (ICI) and the UK Agricultural Research Council had revealed the plant growth inhibiting characteristics of two selective weed killers: one code-named '1313' (Isopropyl Phenyl Carbamate) which it was found resulted in reduced crop yields in wheat, oats, barley, rye and other cereal crops; and another code-named '1414' (Calcium-2-mehyl-4-chloro-phenoxy acetate) which showed inhibiting effects on the growth of sugar beet and other root crops. In a letter to Vannevar Bush, scientific advisor to Roosevelt, Sir John Anderson, an official charged with the organisation of the UKs civilian and economic resources during the War, outlined British research findings as follows:[16]

> The group in charge of the experiments estimate that 1 lb. per acre of either substance would result in almost complete destruction of the vulnerable crops under ideal conditions, and that 3 lbs. per acre would result in almost complete destruction under most conditions that could reasonably be expected.

Initial field trials had shown promising results and Churchill had considered their use against German agricultural targets in 1942, however, no further action was taken at that time. In accordance with Churchill's wishes, research findings pertaining to the above substances had been made available to the Americans, as Anderson put it in a letter to the Prime Minister, 'in case they should wish to use them against the Japanese'.[17]

American progress with regard to anti-crop warfare was relayed back to Churchill during the course of this period.

In his reply to Anderson, which was subsequently copied to the British Prime Minister, Vannevar Bush revealed – with one important exception – the extent to which US work on anti-crop warfare had proceeded along similar lines to British research. According to Bush,[18] US research undertaken was reported to have included:

> (a) Work on plant pathogens involving Late Blight of Potato (*Phytophthora infestans*) and Sclerotium Rot of sugar beets (*Sclerotium rolfsii*). This program was developed by War Research Service and transferred to C.W.S. on October 29, 1943.
>
> (b) A project on organic chemical agents as herbicidals. ... To date one of the most promising compounds appears to be 2-4-dichlorophenoxy acetic acid. [and]
>
> (c) A program is being conducted in Florida by C.W.S. involving defoliation experiments in which large amounts of chemicals are used.

Although UK research findings appeared to suggest a viable military utility regarding anti-crop CW, two obstacles prevented the UK from bringing agents 1313 and 1414 into operational use: first there was a lack of available industrial capacity in the UK which would have prevented such agents being produced in sufficient quantities until 1947; and second UK field trials could not approximate the necessary climatic and soil conditions which would have helped in determining the effect of such substances upon rice and other Japanese rice crops.[19] It is important to note, however, that the British correspondence contained no reference which might have suggested that the UK was interested in anti-crop BW at a level over and above that of basic research.

British documentation dated 12 November 1945, subsequently revealed that US research relating to anti-crop warfare had in fact progressed much further than had been indicated to Churchill in the informal correspondence via Anderson with Vannevar Bush outlined above. A report by the Inter-Services Sub-Committee on Biological Warfare[20] noted that in connection with American plans to employ such agents against Japan and Japanese held islands, some 800 chemical substances, and a variety of biological anti-crop agents had been examined in America as the report states, 'to determine their relative power to injure or destroy plants'. The Annex to the documentation includes an approximation of rice fields in Japan which were calculated to be some 7.9 million acres divided between Japan's main islands (Table 8.1).

Of the 800 chemical substances under consideration in the US only four agents were subsequently selected as the report states, 'for their ability to cause crop injury or destruction at concentrations far lower than common inorganic herbicides.'[21] The chemical agents grouped

Table 8.1 Approximation of rice fields in Japan's main islands

Island name	*Thousands of acres*
Hokkaido	457
Honshu	5895
Shikoku	368
Kyushu	1153
Okinawa	2
Total	7875

under their respective LN designations and their respective target crops are described as follows:[22]

> (a) 2,4 dichlorophenoxyacetic acid (LN8) [subsequently referred to as 2,4,-D], the principal substances employed in the American work and 2 methyl 4 chlorophenoxyacetic acid (LN32) an analogue of equal activity preferred for ease of production in the United Kingdom. These two agents are most active against dicotyleaons, although rice is also extremely sensitive. The remaining cereals are however affected by these agents only at a relatively higher dose level.
>
> (b) 2,4,5 trichlorophenoxyacetic acid (LN14) [or 2,4,5-T], chemically a simple extension of LN8, found to be especially effective in reducing the tuber yield of potatoes, a crop relatively less sensitive to LN8 and LN32.
>
> (c) iso-Propyl N-phenol Carbamate (LN33), especially active against the monocotyledons, i.e. cereals and grasses, but with relatively little effect on broad-leaved plants.

Laboratory experimentation and large field experiments suggested that the above agents 'offered a very promising weapon',[23] and the following approximations would be required for a large scale attack on the areas in Japan identified above: Rice would require 20,000 ton of LN8; Corn, 10,000 of LN33; and Roots, 1,000 ton of LN8 or LN32.

With regard to biological anti-crop agents the report[24] went on to state that:

> '[s]everal fungi known to cause serious disease and destruction of crop plants have been studied in America as possible means of attack of enemy crops. The fungi chiefly considered are:
>
> **Agent C**. *Sclerotium rolfsii sacc.* This attacks a wide range of plants, the most susceptible being herbaceous annuals. Important world crops known to be very susceptible include tobacco, soya beans, sugar beets, sweet potatoes, cotton. The disease produced is typically a stem rot of the seedling, causing death of the plant.
>
> **Agent LO**. *Phytophthora infestans* (Mont) de Bary. The potato is the chief plant of economical importance attacked by this fungus. The disease produced is known as 'late blight' of potatoes.
>
> **Agent IE.** *Piricularia oryzae* Br. and Cav. Some varieties of rice are very susceptible and others resistant to attack by this fungus. The resulting disease has been reported to be serious in Japan, China and India, but it has not been of much importance in the USA.

> **Agent I.** *Helminthosporium oryzae* van Brede de Haan. This fungus causes 'seedling blight' and 'brown spot' on young rice plants and may also cause damage to the maturing head. Crop losses due to this agent occur in Japan, China, Philippines, East Indies and USA.

In further contemplation of an attack on Japanese crops the report[25] went on to state the future potential of such agents:

> The climatic condition necessary to produce an epidemic infection are exacting, but the American authorities believe as a result of laboratory and small scale field experiments that if a very massive dose, as contemplated, was used this method of attack might well give results of value even under average climatic conditions. ... As a method of warfare, therefore, the distribution of plant pathogens is not yet regarded as certain of success, but by establishing many foci of infection and by the use of resistant fungi maintaining their viability under adverse conditions, there seems to be a definite chance of successful attack. The development of more virulent types of fungi must also be considered.

The promising results suggested by the limited testing of the above agents in both the UK, where testing was restricted to chemicals, and the US, where testing with both chemical and biological agents was performed, set the scene for further research and development into the potentialities of such agents in the post-war period. The UK Chiefs of Staff[26] recommended that the following matters receive full consideration:

> (a) the difficulties of defence against such attacks
> (b) the value of these agents for use by a world security organisation as a form of sanction against a recalcitrant nation which would be more speedy than blockade and less repugnant than the atomic bomb.
> (c) their possible use for purposes of internal security within the Empire, e.g. for the destruction of food supplies of dissident tribes in order to control and area; and
> (d) their use for the control of vegetation of ranges, airfield etc.

Indeed there is some evidence which suggests that prior to the end of the war in the Pacific provisions had been made by the US to ship a quantity of anti-crop chemical agents – which along with insect pests were classified, until 1956, as BW agents[27] – to a location in the Pacific close to the Japanese held island of Iwo Jima. The single reference to this claim appeared in a book by Baldwin W. Hanson[28] who wrote in 1950 that:

> in July and August, 1945, a shipload of U.S. biological agents for use in destruction of the Japanese rice crop was en route to the Marianas.

However, in subsequent correspondence Truman confirmed that at that time he had neither approved an amendment to any Presidential Order in force regarding biological weapons, nor had he at any time given approval to its use (see Appendix II).

Aspects of UK anti-crop warfare activity in the post-war period

A post-war UK directive[29] on BW authorised by the Cabinet Defence Committee in 1945 approved the continuation of wartime BW research 'in peace-time'. In commenting upon this directive some five years later the Chairman of the Inter-Service Sub-Committee on Biological Warfare noted[30] that offensive biological warfare research was 'implicit' in this directive. The Cabinet Defence Committee directive was followed by a Defence Research Policy Committee (DRPC) recommendation that, 'research on chemical and biological weapons should be given priority effectively equal to that given to the study of atomic energy.'[31] Indeed, in its 1947 Report[32] on Biological Warfare the UK's Biological Warfare Sub-Committee placed a particular emphasis on offensive biological warfare developments thus:

> On the offensive side [it argued] ... the bias of research and development must be towards the most effective dispersion and utilisation of agents as an airborne cloud, and the development of agents and apparatus required to attain this end.

In regard to the distinction between offensive and defensive preparations the Sub-Committee[33] went on to note that:

> the development of defence measures is so closely linked with offence aspects of the work that no clear division can be made ... the results of any experiments can be used either with intent to develop and efficient B.W. agent, or to furnish a specific preventative against the same agent.

According to Carter and Balmer[34] the period between 1946 and 1951 was characterised by the following seven activities:

- the design and planning of a new establishment adjacent to the Chemical Defence Experimental Establishment;
- the design and erection of a pilot plant to study the basics of large-scale production of bacteria;
- an expansion of fundamental research;
- the recruitment and training of new staff;
- the mounting of a major biological warfare trial, Operation Harness, at sea, off the Leeward Islands;
- the devolvement to the Chemical Defence Experimental Establishment of research and development for biological warfare munitions and weapon design; and
- the setting up of a new and eminent body of independent advice, the Biological Research Advisory Board.

A further Cabinet Defence Committee Meeting implicitly approved offensive anti-crop BW research and development in the UK as part of the following directive:[35]

> to protect the Civil population and Service personnel, as well as crops and livestock against attack by biological methods and to retaliate by these methods against the enemy should the Government of the day decide to adopt this course.

In a climate of post-war austerity[36]offensive UK BW research was afforded a high priority between the end of the War and 1950. The following elements of the offensive programme were characteristic of the priority given to the programme during the course of this period. First, there existed a perception in the UK that the country was potentially vulnerable to a BW attack with the proliferation of such weapons resulting in a possible enemy BW capability by 1951.[37] Secondly, although it was not considered possible to achieve such a capability prior to 1957, a requirement for a strategic offensive anti-personnel BW bomb had been lodged with the Ministry of Supply by the Air Staff in 1946.[38] The requirement stated that production of this munition would be achieved by 1951 resulting in a DRPC recommendation to increase the level of research and development on BW munitions. And thirdly, this period saw considerable financial investment resulting in the construction of a new building, which opened in 1951, for the Microbial Research Department at Porton Down.[39]

The DRPC's Review of Defence[40] in 1954, however, resulted in a low priority being assigned to offensive BW research and although the Cabinet Defence Committee had endorsed a directive which placed emphasis on preparations that included a retaliatory BW capability,[41] BW research continued but was reported to be mainly defensive by 1955.[42] Although no clear policy on offensive UK BW emerged between 1953 and 1958, according to Carter and Balmer[43] it seemed most likely that UK BW was 'eventually subsumed' under a secret Cabinet decision abandoning the UK's offensive chemical warfare capability in 1956. The demise of the UK offensive biological weapons programme can be traced to its end in 1961.[44]

No clear statements pertaining to the division of labour on matters relating to UK and US offensive anti-crop BW research have thus far emerged into the public domain.[45] Nevertheless, agreement of allocation of research and development on matters relating to BW was reached during the course of annual meetings between UK and US BW workers. According to the Chief Scientist, at the UK's Ministry of Supply:[46]

> [a]s an example of this co-operation, agreement was reached at a meeting held last year that work on some 25 BW or CW projects should be undertaken by one or other of these countries. Particularly in the CW and BW fields the exchange of information and samples (through the medium of liaison officers at the Research Establishments on both sides of the Atlantic) is very considerable.

In the immediate post-war period, however, it was possible to draw a distinction between offensive BW research in the UK, and the US. What emerges from the available sources[47] is a UK commitment to the pursuit of basic offensive BW research and a long-term strategy toward an effective offensive BW capability. Whereas in the US there existed a commitment to the bulk production of agents and short-term developments relating to the acquisition of a retaliatory capability.[48] According to the Ministry of Supply:

> with regard to biological warfare, basic research is to be carried on primarily to identify and estimate the threat against us and secondarily, and then in the long term, with a view to offensive use of biological agents. At present we are relying on the United States for the short term development of biological weapons.

The formulation of the above directive in the UK applied to developments relating to anti-personnel and anti-crop BW. Throughout this period, however, the UK's commitment to basic research on offensive BW began to be brought into question and this matter was the subject of periodic reviews until the early 1960s,[49] with the UK coming to rely more and more on the US for the production of offensive biological weapons.[50]

Further declassified British documentation offers an insight into the level of anti-crop transatlantic co-operation between the end of the War and the latter half of the 1950s. In the post-war period, the UKs Crop Committee had been set up by the Ministry of Supply in 1948 with a remit to, 'look into methods of destroying crops and other vegetation and countermeasures'[51] under the auspices of the Advisory Council on Scientific Research and Technical Development.

The Ministry of Supply recommendation[52] to set up a scientific body to report on crop destruction contained an interesting appraisal of the state of play regarding anti-crop warfare on both sides of the Atlantic in 1948:

> At present, the subject of crop destruction is in a very undeveloped state. During the war some progress was made, principally in U.S.A., in the warlike applications of chemical agents, but the quantities raised formidable production and tactical problems. Less progress was made in connexion with the use of biological agents, though these may in the end prove more effective than chemical agents.
>
> Since the end of the War, work on a limited scale has been going on in U.S.A. and the Americans are now inclined to think that plant pathogens offer the greatest potentialities. In this country, no work is being done other than research on weed killers for agricultural purposes and University research of a fundamental nature in plant pathology.

At the first meeting of the Crop Committee,[53] 10 August 1948, Professor F.T. Brookes, Chairman of the Committee, reviewed possible methods of crop destruction by both chemical and biological means revealing that chemical weed killers were being actively studied in the UK, and perhaps most interestingly, the limited extent to which anti-crop BW had received attention by UK researchers in the immediate post-war period. In this connection, recent experiments, conducted at Cambridge University, had highlighted the difficulties associated with establishing yellow rust epidemics in wheat crops and had shown that anti-crop BW was in practice quite a difficult undertaking. The Second Report[54] of the Crop Committee submitted some months later in November 1948 reported further in connection with UK anti-crop BW research and developments that, 'no work of interest to the Committee was being carried out' at the Microbiological Research Department, at Porton Down.

The Crop Committee, and the committee that subsequently replaced it in the mid-1950s, reported to the Advisory Council on Scientific Research and Technical Development at six-monthly intervals and between 1948 and 1955 had conducted a wide range of investigations into crop warfare and defensive measures. Investigations conducted by the Crop Committee, for example, included: herbicide research (1948);[55] investigations into the water content of cereals (1948);[56] spore suspension research (1949);[57] investigations into aircraft in agriculture (1949);[58] investigations into anti-crop CW (1949);[59] investigations into countermeasures to anti-crop and anti-animal warfare (1949);[60] investigations involving 2-4,D (1950);[61] investigations into phytotoxicity (1950);[62] investigations into the effects of insecticides and herbicides on animals (1951);[63] investigations into anti-crop detection and destruction of destructive agents (1952);[64] and, anti-crop aerial spray trials (1954).[65] Further reports relating to investigations into anti-crop warfare were also conducted during the course of this period including: investigations into analysis of rice blast epidemic caused by *Piricularia oryzae* (1954);[66] a report on cereal rust;[67] the screening of *Piricularia* isolates for pathogenicity to rice (1955);[68] and, investigations into clandestine attacks on crops and livestock (1955).[69]

From its inception, the Crop Committee received regular updates with regard to American progress on anti-crop warfare and was in receipt of: US Chemical Corps *Quarterly Reports* (1948[70] and 1949[71]); US Chemical Corps Report[72] and Camp Detrick *Technical Reports* (1950);[73] Camp Detrick Report on plant inhibitors on sugar beet crops (1952);[74] Extracts from US Chemical Corp Research and Development Quarterly Progress Report (1953);[75] secret studies in drying BW agents (1953);[76] a report on work of the Crop Division at Camp Detrick (1955);[77] and, a report on the processing, packaging and storage of cereal rust spores (1955).[78] However, anti-crop warfare information-sharing arrangements would begin to falter in the latter half of the 1950s with reports on US progress ceasing completely toward the end of the decade.

In the mid-1950s the UK Agricultural Defence Advisory Committee took over as the constituent branch of the Advisory Council on Scientific Research and Technical Development concerned with anti-crop warfare with a wider remit than that of the Crop Committee. The terms of reference mandated the newly formed Agricultural Defence Advisory Committee[79] to:

> review current research which may relate to possible methods of destroying or decreasing the effective yield of naturally produced foodstuffs or raw materials, with particular reference to new developments which appear likely to make a significant change in the potentialities of this type of warfare. ... These terms are wider than those of the previous Crop Committee and cover potential methods of enemy attack on vegetable crops and livestock in this country [the UK] and in Colonial territory ... and appropriate defence measures.

Based upon investigations carried out by the Crop Committee, the work of the Agricultural Defence Advisory Committee fell into three main classes:[80] 1. the effects of radio-active fission products on plants and animals; 2. the potentialities of bacteriological warfare for the spread of animal disease both in this country and in the Colonies; and 3. the potentialities of crop warfare by chemical and biological means.

Although American anti-crop warfare investigations were carried out on a much larger scale than those conducted in the UK, chemical anti-crop agents were first deployed by the British. Testing with chemical agents 2,4-D, 2,4,5-T, and with endothal (3,6-endoxohexahydrophthalic acid), took place under tropical conditions in Tanganyika, and 2,4,5-T was used in Kenya during a programme to eradicate tsetse fly by aerial spraying.

The first military application of British anti-crop chemicals took place in Malaya. During the course of 14 year Malayan conflict, *n*-butyl ester of 2,4,5-trichlorophenoxyacetate, was used in a programme of roadside defoliation to thin jungle cover and subsequently reduce the likelihood of ambush. Although the roadside defoliation programme was restricted to surface spraying, anti-crop chemicals were also dropped from helicopters as part of a related food denial programme, a tactic considered by the general commanding officer in Malaya as one of the 'decisive weapons' in the British-supported campaign against Malayan insurgents. According to one source, British deployment of chemical herbicides in Malaya formed, 'the basis for future American involvement in herbicidal warfare in Vietnam'.[81] British deployment of anti-crop chemical agents in Malaya, however, are thought to have had little impact on events in Korea[82] where anti-crop CW agents were not used.

Although US Chemical Corps funding had reduced in accordance with overall reductions in military spending at the end of the War, anti-crop warfare research and development had continued in the post-war period. Between 1943 and 1950 some 12,000 chemical agents had been screened for their potential as anti-crop chemical agents. With the defoliation programme operating at a reduced level, between 700 and 1000 chemical compounds were subsequently selected for further research and development.

During the course of the war in Korea, butyl esters of 2,4-D and 2,4,5-T, were selected in the US for investigations which began with the MC-1 'Hourglass' bulk dissemination anti-crop agent spray system installed on a variety of aircraft. However, according to one source the, 'Korean War ended before the system was used, and the equipment and chemicals were placed in storage.'[83] According to Cecil[84] the above agents were not used again until 1959 when they were brought out of storage and pure butyl 2,4-D and butyl 2,4,5-T was mixed and applied in a one-to-one ratio during a defoliation programme on a firing range at Camp Drum, New York. This programme, he argues, was the, 'first large-scale attempt at airborne military defoliation [by the US].'

A UK cereal rust epidemic – an unusual and suspicious outbreak?

UK research into BW against crops between 1948 and the late 1950s appears not to have progressed much further than laboratory research with specific organisms affecting certain cereals. As noted above, such laboratory work included investigations into pathogens that cause cereal rust, and disease in rice, and this aspect of the UK programme appears not to have been expanded during the latter half of the 1950s.

The occurrence of a minor crop disease epidemic in the UK in 1955 had provoked considerable US interest. The cool wet spring and the subsequent fine warm summer of that year had resulted in an unusual, and serious, outbreak of black rust disease on wheat in south-west England. While the climatic conditions favourable to the outbreak of this crop disease were good, such conditions were also favourable to the growth of wheat, and the wheat yield in that region in 1955 had been exceptional. The exacting climatic conditions for the epidemic spread of this crop disease, however, were seldom favourable in the UK, and the black rust outbreak had prompted media and scientific interest in both countries.

In response to press reports of the disease outbreak the Americans had filed a request for further information with the UK representative of the British Joint Services Mission in Washington. After the initial exchange of correspondence concerning the black rust outbreak a request came from the US Chemical Corps for further information on: 1. the varietal behaviour of the rust disease (i.e. which wheat varieties it attacked); 2. the strain of the pathogen and the resistance of the wheat varieties in question; and, 3. the source of the pathogen (i.e. where it came from).

Concerning these requests, the minutes[85] of the Fourth Meeting of the Agricultural Advisory Committee in March 1956, noted that information relating to the varietal resistance of the crop was in the process of being accumulated by the UK Agricultural Advisory Service, and that an undertaking had been received from both France and Portugal who had promised help with information in respect of the source of the pathogen. Further investigation was also underway in relation to the possible source of the outbreak by examining the possible correlation between the black rust outbreak and the large number of airborne rust spores collected on slides during an unrelated investigation into allergies in Bristol.

As black rust disease had not been considered to be of sufficient importance to justify more detailed investigation in the UK, only limited information on the strains of this disease had been accumulated by the Agricultural Advisory Service and it was agreed that action would be taken through the UK Ministry of Agriculture for samples of the spores to be sent to the US for identification. That the UK lacked the means by which to identify the pathogen in question is perhaps indicative of the low level of research and development on plant pathogens as anti-crop warfare agents in the UK during the course of the post-war period. Nevertheless, it will be noted that British developments in anti-crop BW were not inconsistent with the stages in acquiring an operational BW capability as identified by the US Office of Technology Assessment (OTA) in Table 1 (Chapter 1, see Stages 1 to 3).

Aspects of US anti-crop warfare activity in the post-war period

The scope of US anti-crop warfare activities in the 1950s was outlined in a report[86] by the UK's Agricultural Defence Advisory Committee in December 1955. Some three years previously, a UK representative[87] had noted with some alarm that the Americans had developed a 'most aggressive outlook', possibly related to the desire to bring the war in Korea to a conclusion, and that US BW had shifted to short-term research and development on anti-personnel and anti-crop weapons. Entitled 'A Note on the Recent Work of the Crops Division at Camp

Detrick', the report was based on the Proceedings of a Technical Staff Meeting on Crops that had been held at Camp Detrick, 14 November 1955. This meeting had been attended by representatives from the UK. US anti-crop warfare activities, it reported, involved: investigations into both chemical compounds and plant pathogens as anti-crop warfare agents; problems associated with the bulk processing, packaging and storage of the plant pathogens under development; and, the development and testing of a number of weapons systems for the dissemination of the above agents (developments relating to munitions are dealt with in chapter 9).

The British documentation recounts the programme of the meeting which included four short discourses on the anti-crop warfare activities of the constituent branches of the Chemical Corps' Crops Division: the Chemistry Branch; the Biology Branch; the Plant Physiology Branch; and the Operational Requirements Branch. The programme of the meeting also included two short films. Most interestingly, one of the films, entitled, *Packaged Famine*, according to the report:

> was in the form of a discourse by Dr. Leroy D. Fothergill, Scientific Advisor to the Chemical Corps Biological Laboratories. This film also covered some of the ground covered in the lectures, but was mainly of interest for its appraisal of the possibilities of employing anti-crop warfare against the Soviet Union.

Further US documentation[88] also reveals that considerable thought had been given to the requirements associated with large-scale chemical and anti-crop BW attacks on the principal rice-growing regions in China.

The objective of the constituent branch of the US Chemical Corps responsible for military investigations relating to chemical anti-crop warfare agents is stated in the British documentation as follows:

> The principal objective of the Chemistry Branch is, as always, the development of a universal anti-crop agent, i.e., one which would be equally effective in reducing the yield of narrow-leaved as well as broad-leaved crops at very low concentrations.

The Chemistry Branch of the Crop Division was capable of conducting preliminary investigations into new chemical compounds as potential anti-crop warfare agents at a rate of approximately 200 per month. Investigations had revealed that butyl 2:4 dichlorophenoxyacetate (LNA or 2,4-D), and butyl 2:4:5-trichlorophenoxyacetate (LNB, or

Table 8.2 The effect of KF and LNA on different crops

Crop	*Yield reduction (%)*	
	KF	*LNA*
Rice	100	2
Oats	93	13
Millet	86	20
Cabbage	14	98

2,4,5-T) were very effective against broad-leaved crops such as beans, cotton, sugar beet, and sunflowers. These agents proved to be less effective, however, against narrow-leaved plants such as oats, millet, rice, rye, and wheat. Conversely, investigations conducted with a further chemical agent (4-fluorophenoxyacetic acid) known as 'KF' had proved that this agent was very effective against narrow-leaved plants but less so against broad leaved plants (Table 8.2).

As chemical compound KF was a solid, a number of problems emerged that made application of this agent problematic. KF was found too difficult to disperse from a munition and its composition made absorption into target plants slow. However, investigations relating to a number of liquid derivatives of KF had revealed up to 92 per cent effectiveness of the parent compound when tested for activity towards both rice and millet. Whilst it was reported that a number of other chemical compounds had aroused interest during the course of investigations by the Chemistry Branch of the US Crop Division, the derivative of KF represented the 'principal candidate'[89] chemical agent for the attack of narrow-leaved crops in 1955. Related investigations conducted during the course of research by the Physiology Branch of the US Crop Division also revealed that the KF derivative was particularly active against wheat. At this time, however, no single chemical anti-crop agent had been identified that was effective against both broad- and narrow-leaved crops. According to the British documentation in 1955:[90]

> The search for the universal anti-crop agent may still require many years of research, but the discovery of the activity of KF against narrow-leaved crops does mean that it is now possible to provide agents capable of attacking the major food crops of the world. It may well be that a mixture of LNA (or LNB) and a suitable derivative of KF will serve, to all intents and purposed, as a universal anti-crop agent.

The rate of testing of such agents was expected to increase once the construction of new laboratories and greenhouses was completed in early 1956. More comprehensive testing of such agents was also expected to proceed with further development of testing grounds at Gallatin Valley, Montana (for wheat), and at Beaumont, Texas (for rice). Limited investigations into the action of Defoliants and Desiccants were also conducted by the Chemistry Branch of the US Crop Division during this period.

Defoliants and desiccants

According to the British documentation the military utility of defoliants and desiccants was stated as follows:

> defoliants are of potential military value in jungle warfare in the prevention of ambushes, the exposure of enemy strong-points, and the improvement of friendly fields of fire.

The following defoliants were the subject of investigation during the course of this period and are presented in order of effectiveness:

(a) Endothal-sodium 3:6-endox-hexahydrophthalate
(b) 2-Butyne-1:4-diol
(c) Magnesium chlorate

The following desiccants are also presented in order of effectiveness, thus:

(a) Pentachlorophenol
(b) Tributyl phosphate
(c) Dinitrophenols

However, it was reported that no further experimental work had been conducted with regard desiccants listed above.

Plant pathogens

Investigations into plant pathogens as anti-crop warfare agents were conducted under the auspices of the Biology Branch of the US Crop Division during the course of this period and a major part of the work was devoted to the problems associated with production and storage of such agents. The following plant pathogens were identified in the British documentation as, 'those that offered the greatest potential in the attack of food crops':

(a) Stem rusts of wheat and rye (*Puccinia graminis*)
(b) Blast of rice (*Piricularia oryzae*)
(c) Late blight of potatoes (*Phytophthora infestans*)

Each of the above plant pathogens will be discussed in more detail in relation to production problems associated with: (a). the preparation of agents for drying; (b). the drying process; (c). the storage stability of the pathogens; (e). distribution problems; and (f). the promise of dry agent for military use. It will be noted that the pathogens discussed in the following section are those that have been identified as the causal-agents of losses to crops of major economic and social significance (see Chapter 3). It will be recalled also that such agents have also been identified during the process of negotiating a compliance Protocol for the 1972 (BTWC/ as Pathogens of Importance to the BTWC (see Chapter 4). Furthermore, *Puccinia graminis tritici* and *Piricularia oryzae* were identified as agents that were subsequently standardised by the US under the military designations of 'TX' and 'LX' respectively and assimilated in the US strategic war-fighting arsenal.[91] Unfortunately, there are no references available in the open literature which confirm the standardisation of the causal-agent of late blight of potatoes during the course of the US anti-crop BW programme.

Stem rusts of wheat and rye

The causal agent of stem rust of wheat (*Puccinia graminis*) had long been identified in the US for its utility as a potential anti-crop BW agent. Research had begun with *Puccinia graminis* in the mid-1940s and by 1955 the large-scale production of this pathogen was underway.

The basis for adopting certain strains or races of this pathogen for military purposes is described in a US study entitled *Secret Studies in Drying B.W. Agents*[92], thus:

> The races which would be used in a BW program would be races which would not infect the commercial varieties in this country but would attack those grown in the target areas.

Preparation of agents for drying

For military purposes the urediospore[93] of *Puccinia graminis tritici* is the normal disseminating form of this organism. The production of this organism, which in relation to its plant host is described in phytopathological terms as an 'obligate parasite',[94] begins in fields planted with grain.

In order to produce large quantities of this agent growing fields of grain are infected with the organism and spores are repeatedly harvested from contaminated plants once the local epidemic has progressed to a suitable degree of severity. In relation to its physical characteristics this organism measures some 15 by 25 microns, has a bulk density of 0.5 and a mass density of between 0.8 and 1.0. Extraneous debris, such as particles of soil or dirt and other sources of contamination, are then separated from the harvested spores which are kept at a low temperature (ca 45F) in preparation for the drying process.

Further British documentation[95] which included an extract from a Camp Detrick Special Report on processing, packaging and storage of cereal rust spores highlighted some of the problems associated with the large-scale production of the above agent:

> with the adoption of cereal rusts as anti-crop agents and the resultant large-scale production of spores, the development of adequate methods of preserving spore viability and consequent infectivity for the longest possible time became necessary and studies were initiated to accomplish this end.

Investigations into the long-term preservation of viable cereal rust spores lead to the standardisation of a method of maintaining the moisture content of spores and thus achieving a storage half-life of approximately 9–12 months. This method, which involved the utilisation of industrial vacuum-drying equipment, could be used for both bulk- and small-scale production of cereal rust spores.

Further investigation into extending the storage half-life of cereal rust spores revealed that small quantities (10–100mg) of the pathogen could remain viable for between 2 and 3 years but would subsequently require hydration treatment in order to induce germination. During the course of this period, according the above report:[96]

> Efforts [were] being made to adapt this method of processing to the large quantities involved in field production.

The production of the causal agent of stem rust of rye had not been possible on a scale comparable with wheat stem rust at this time due adverse weather conditions at the sites chosen for the production of this agent (Nebraska, South Dakota, and Idaho). According to the above report, production of this agent on a large-scale, however, was considered 'feasible'.

The drying process

Two methods had been shown to be effective for the drying of large quantities of rust spores. One technique, known as the fluidised bed technique involved the use of a powder conditioner that had been designed and built by Camp Detrick's E Division, was favoured for its relatively short drying time period but its special design and the cumulative losses of agent when a series of production runs had been made meant that a second technique was favoured. This method utilised a standard vacuum tray drier. Harvested spores were placed in the trays in the vacuum oven and dried at a vacuum of 29 inches and at a temperature of 120°F, until the moisture content of the spores was reduced to approximately 11 per cent. The harvested and dried spores are then placed in sealed nitrogen-filled cans in 0.5 or 1 lb quantities until ready for use. According to the report, a production facility in Florida was capable of producing in the region of 60–70 lbs, of viable *Puccinia graminis* spores per acre.

The daily production capacity of a six-tray oven measuring 24 × 36 ft loaded with approximately 52 lb of spores had a daily capacity of approximately 420 lb of dried spores. When stored in such a manner (as a free-flowing powder) with a moisture content of 10–12 per cent the spores were found to be between 80 and 90 per cent viable.

Storage stability

Table 8.3 demonstrates the storage stability of *Puccinia graminis* (Oat Stem Rust spores, Race 8) when stored under various conditions. This

Table 8.3 Viability of spores of oat stem rust, race 8, after storage under different atmospheres (% germination)

Treatment	*Months of storage*								
	0	*3*	*6*	*9.5*	*13*	*16*	*18*	*20*	*22*
Unsealed: 50% RH	71	15	10	5	0	3	0	0	0
Sealed under air	71	44	40	31	12	0	22	0	0
Sealed under nitrogen	71	47	45	61	58	51	54	41	24
Sealed under helium	71	58	43	60	62	46	36	19	20
Sealed under vacuum	71	48	39	64	65	57	57	44	31

investigation demonstrated that the longevity and viability of rust spores could be significantly extended when spores were either stored in a vacuum, or under nitrogen or other inert gases with over 30 per cent of the spores remaining viable for periods of up to 22 months.

Further investigation revealed that the longevity and viability of spores reduced significantly when stored together with the feather agent carrier (Figure 8.1).

Distribution

Over a target area representing some several hundred square miles dried spores would fall or be disseminated over the target area by air

Figure 8.1 Viability of processed spores stored on bulk carriers over a period of six months[97]

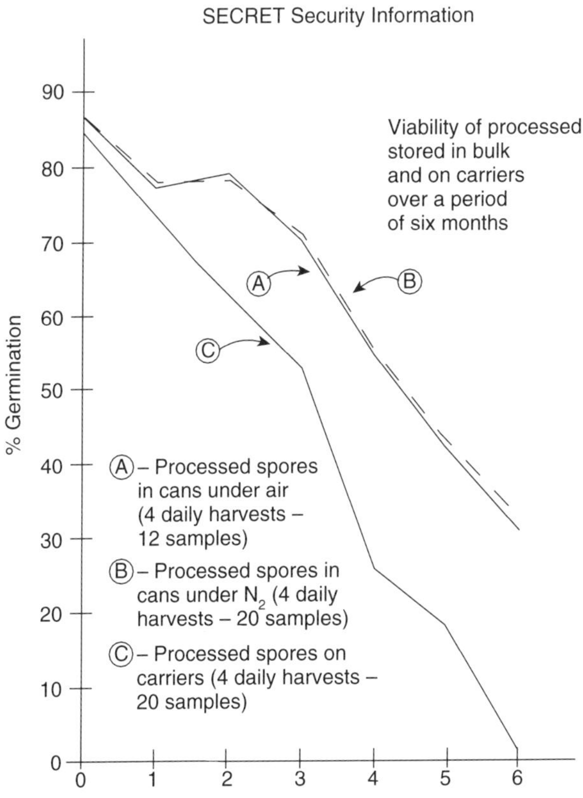

Source: 'Secret Studies in Drying BW Agents,' *Camp Detrick Techical Bulletin*, No. 4, Crop Committee, AC12368, at PRO, WO195, 12364.

currents. Where the spores remained dry prior to their dissemination among the host crops viability remained high. However, viability decreased where spores had come into contact with water prior to their arrival on target. Both water and high winds were found to have the effect of reducing infection by mechanically removing the agent from the plant. Favoured conditions for successful infection were found to be where a clear sky and a light dew formation on the host plant at night following spore deposition. The report[98] describes the way in which infection occurs under ideal conditions thus:

> The period in which dew (free moisture) is needed on the plants to initiate infection is roughly twelve hours. This represents the period required for the spore to germinate and for the germ tubes to penetrate through the stomates and initiate infection in the leaf tissue. At this time the fungus becomes independent of the spore and obtains its nutrient from the host tissue. Once this has occurred, external environmental conditions will not affect the persistence of the agent but do have an effect on the development of the agent in the host tissue and the subsequent production of spores on these plants.

The promise of dry agent for military use

As noted in the report US anti-crop BW workers needed to look no further than standard academic texts for evidence of the effectiveness of this agent with respect to the destruction of crops. However, it was noted in the report that many agronomic problems needed to be overcome prior to production of *Puccinia graminis* being placed on a certain basis. At this time work on strains of *Puccinia graminis* effecting wheat were at a more advanced stage than work with rye and oat stem rusts.

Blast of rice

The normal disseminating form of *Piricularia oryzae* (the causal agent of 'rice blast') and *Helminthosporium oryzae* (the causal agent of 'seedling blight' or 'brown spot' of rice) is for military purposes by means of a stable spore. This is a hardy spore that is resistant to drying and adverse weather conditions. Mycelial fragments of this organism when mixed with a carrier or dilutent can also be disseminated to cause disease in crops. As these agents are not obligate parasites production of these agents took place on cultures under controlled factory conditions.

Preparation of agents for drying (agent material containing spores)

The strains of these agents[99] were selected in accordance with two criteria: 1. the strain's pathogenicity; and 2. the ability of the agent to sporulate on rice polish agar and oats–sorghum substrate.

The above agents were grown on steeped corn. Eleven per cent of the weight of the corn was made up by water. The steeped corn was spread to a depth of 1.5 inches on trays measuring 20 × 30 × 3 feet and covered with a lid. The lid contained a perforated coil containing a spore suspension and the enclosed grain was aerated. In this way inoculation of the grain was achieved. The inoculated grain would then be aerated in sterile, humidified air and incubated at 25°C in a constant temperature cabinet for up to 10 days until maximum inoculation of the spores per gram of dry weight was obtained.

The spore-laden grain was then placed in a cabinet dryer at a temperature of 40°C for a period of 48 hours. This process produced a total weight of spore-laden grain of 33 lb 12 oz. The grain was then divided into three lots and processed with the following three solvents: carbon tetrachloride; ethyl acetate; and, isopropyl alcohol. Each lot was placed in a rotating drum with each respective solvent and agitated for a period of 5 minutes. The suspension was then drained off and fresh solvent was applied. The suspension was then combined with 500 g of celite, and 1500 g of peat. Each suspension was then filtered and placed in a cabinet dryer for a period of one hour.

Table 8.4 shows the production amounts of spores per gram of suspension with each of the above solvents.

In preparing spores for the drying process the above method resulted in an entire production run of 5,222 g of substrate containing 190×10^9 spores.

Table 8.4 Amounts of spores per gram of suspension in different solvents

Weight of lot (sporulated grain)	*Solvent used*	*Filler (peat and celite added) (g)*	*Dried product recovered (g)*	*Final product spores / g ($\times 10^6$)*
11 lb 4 oz	Carbon tetrachloride	2000	1697	36.5 (= 36,500,000 spores per g)
11 lb 2 oz	Ethyl acetate	2000	1742	41.8
11 lb 6 oz	Isopropyl alcohol	2000	1783	31.0

Drying process for agent material containing spores

With the spore/grain material placed in a cabinet dryer the moisture content of the spores and grain could be reduced by circulating air heated to 40°C over the material for a period of 24 hours. This reduced the moisture content of the material to between 7 per cent and 10 per cent. The spores would then be removed from the grain by mixing the material with a non-aqueous solvent and placing the material/solvent mix in a rotating agitation drum. In order to reduce the accumulation of residual solvent, fillers (such as peat) and preservatives (such as celite) were added to the material/solvent mix and the spores were then removed from the resultant suspension by filtration. The filtered produce was then placed in a cabinet dryer for a period of up to three hours.

Thus one three compartment dryer, fitted with 18 trays measuring 20 × 30 × 3 feet could dry up to 423 lb of sporulated grain during the course of a 24-hour production cycle. Full utilisation of available production facilities (identified in the report as T-325) resulted in the following production estimate:[100]

> By using the peat and celite fillers at a rate of 1 gram per 50,000,000 spores, approximately 115 lbs of dry spore, peat, celite material could be produced per day.

After the drying process, the resultant spore, peat and celite material had a moisture content of between 7 per cent and 9 per cent by weight. While the particle size of the filler and preservative varied considerably this production process resulted in relatively uniform spore size measuring between 10 and 20 microns. Stored under completely dry conditions at a temperature of 8°C the dried product was shown to maintain viability for over two years.[101] Although data on the relationship between viability and infectivity was not available at this time it was reported that the presence of filler and preservative would not reduce the viability of the spores.

When tested for viability in 1951 miscellaneous collections of spores (and mycelial fragments) that had remained in storage at a temperature of 5°C and at 10 per cent to 20 per cent relative humidity yielded the following results (Table 8.5).

Distribution

Although no data on the behaviour of the above agent as an aerosol was available at this time, distribution methods applied to the rusts were considered to be applicable for the anti-rice agents discussed

Table 8.5 Viability of spores tested in 1951

Organism	*Date prepared*	*Germination (%)*
Piricularia oryzae	July 1943	75
"	June 1945	30
"	July 1945	75
"	July 1945	50
"	July 1945	50
"	July 1945	30
"	June 1947	80
"	July 1947	90
"	July 1947	80
"	July 1948	80
Helminthosporium oryzae	March 1943	90
"	February 1949	50

above. Like the rusts the spores of the anti-rice agents were transported to new sites from infected plants by air currents. Once deposited on the target plant adequate surface moisture would be required in order that germination could take place. Sufficient moisture must remain on the plant for a period of between 10 and 16 hours to allow time for penetration of the leaf tissue. Optimum temperature range for the germination of the above organisms was between 20°C and 30°C.

The persistence of the anti-rice agents would be reduced, however, by the insufficient presence of moisture on the leaves of target plants. Spores could also be mechanically removed from the host plants by wind and rain but apart from its effect on moisture, sunlight was reported to have little effect on the organisms persistence.

Promise of dry agents for military use

Due to the stability of the above anti-rice agents in storage it was though conceivable that large quantities could be produced and placed in storage until required for use. Under the right conditions anti-rice agents were thought to be particularly effective. According to the report:[102]

> This group of organisms can cause a great amount of damage. ... There is no reason to believe that a reduction in crop yield cannot be achieved through the use of these organisms. The extent of loss and the military feasibility will depend on the methods of distribution and the accuracy with which meteorological conditions can be forecast.

Late blight of potatoes

The normal disseminating form of the causal agent of late blight fungus (*Phytophthora infestans*) in nature is by means of a fragile thin-walled spore called a 'sporangium'. However, for military purposes production of this agent concentrated on the development of pellets of porous material that had been infested with the fungus. The following explanation is presented in the report in relation to the adoption of this method of agent production:[103]

> The assumption is made that these pellets would provide an initial source of inoculum for field infection following their distribution in the field and exposure to conditions of high moisture and low temperatures.

Preparation of agents for drying

The strains of this pathogen were selected for production at this time were those that had been shown to attack the most resistant varieties of potato grown commercially in the US in the early 1950s. Like the anti-rice pathogen this organism grows readily on artificial media and large quantities of sporangia could be grown on a mixture of four-parts peanut hulls and one-part grain. The addition of small quantities of metallic iron was shown to have the effect of increasing sporulation four-fold. After a period of 10 days after inoculation of the media this method was shown to yield 1,500,000 sporangia per gram of substrate.

As sporangia of *Phytophthora infestans* would lose viability in air at relative humidities of less than 100 per cent development of the sporangia of this organism as a dry agent was ruled out.

According to the report the pellets of porous material used to disseminate this organism were prepared as follows:[104]

> 4 lbs dehydrated potatoes, 5 lbs ground brown-black peat, 4 lbs ground peanut hulls, and 3 eggs per lb were mixed with sufficient water to make a mass that was moist but yet stiff enough to retain its shape when moulded into a ball. The mixture was made into a uniform consistency by mixing in a Unique Ribbon Mixer. This 'dough' was spread in a large greased (lard) baking pan to a depth of one inch. It was then heated in an autoclave for 20 min at 250°F to coagulate the egg albumin.

The above preparation was then cut into 1 inch cubes and dried at a temperature of 130°F for a period of approximately 24 hours.

During inoculation the pellets were placed in culture vessels and saturated with water. The inoculum consisted of a water suspension of sporangia with at least 25 sporangia per 16 mm field. A few cubic centilitres of the suspension was sprayed from a vessel containing the sporangia over the pellets through a 14 gauge hypodermic needle. The culture vessels were then sealed and stored at 20°C for incubation resulting in sporangia being produced in abundance on both the inside and outside of the pellet.

The drying process

A six-tray vacuum tray drier supplied by the Stokes Machine Company proved to the most effective means by which pellets could be dried. The wet pellets were placed in the trays which measured 24 × 36 × 36 inches and dried for a period of 2 hours at a vacuum of 28.5 inches and a temperature of 186°F. This process produced a one inch pellet with a dry shell and a moist interior. Further investigations with regard to the infectivity of the pellets revealed no loss of virulence resulted from the drying process.

Storage stability

A number of factors were found to influence the viability of the organism in the pellets in storage and in order to maintain the viability of the organism the pellets required a constant temperature at between 5°C and 20°C, an environment free from contamination, and a residual moisture level within the pellet.

Dissemination

Due to the necessity of developing the pellet agent carrier for *Phytophthora infestans*, the production of this organism in dry/dust form was, from an operational standpoint, ruled-out. Further investigations into the aerosolisation of this agent had also revealed that when suspended in a 25 per cent dextrose solution sporangia of this organism maintained only a 30 per cent viability for a period of only 20 days when stored at a temperature of 2°C. Although no large-scale dissemination tests had been conducted with this agent it was reported, however, that this agent would be disseminated in pellet form over the target area.

According to the report,[105] after dissemination of this agent the following results might be expected:

> *Phytophthora infestans* is a pathogen capable of rapid multiplication under favourable field conditions. The period between infection and sporulation is less than one week, under very favourable conditions, therefore it would be possible to have 100% infection in a field in one month after inoculum was introduced to a single plant. While heavy application of inoculum would decrease the time for the disease to develop epidemic proportions, scattered primary inoculation would be effective if favourable weather persisted.

Once released into the environment further factors would, however, influence the survival of the pathogen. Fluctuations in temperature and relative humidity, and excessive rain and sunlight could all act to reduce the viability of the pathogen to a matter of several days making difficult accurate predictions of the life of the pathogen in the pellet.

Promise of dry agents for military use

The development of the porous pellet for *Phytophthora infestans* was considered to be a less than satisfactory means of dissemination[106] and successful application of this agent to a target crop would require accurate weather forecasting in order that dissemination could be timed to coincide with the start of rainy periods that could be predicted to last more than one day. Although a number of problems associated with the further development of this agent would need to be overcome, the chances of success with the agent in pellet form were reported to be 'good'.[107]

Divergence

The above investigation into Anglo-American anti-crop warfare collaboration reveals a clear overview of the activities in each respective country and provides an insight into the level of collaboration and the extent to which information was shared on this topic between the mid-1940s and the latter half of the 1950s. In the UK emphasis was placed upon the development of anti-crop warfare chemicals and close examination of the recently declassified British documentation cited above reveals little evidence that anti-crop warfare with plant pathogens was pursued with vigour by the British during the course of this period. Research in this connection, however, did take place but was limited to that of a fundamental nature.

Although the US went on the use chemical anti-crop agent in large quantities in South-East Asia, the Minutes of the Fourth Meeting of the Agricultural Defence Advisory Committee,[108] in March 1956, included a report on American progress on anti-crop warfare which stated that emphasis was being placed more and more upon the development and dissemination of fungal plant pathogens which might be used to attack the great grain belt of the USSR where large-scale production of short wheats was concentrated. According to American estimates, if directly distributed from aeroplanes, such methods could:

> cripple the USSR in eighteen months by ... dissemination of the rust disease in this grain belt.

Enthusiasm for the deployment of fungal plant pathogens as weapons of war among members of the UK's Agricultural Defence Advisory Committee, however, did not match that of the Americans. It was reported to the Committee that climatic conditions in the grain growing regions in question were not favourable to the epidemic spread of rust disease and the proceedings of the Fourth Meeting of the Committee were brought to a close by its Chairman who declared that this form of warfare, '... seemed to be a complicated way of arriving at a doubtful result'.

By 1957, reports of American progress on anti-crop chemical and biological warfare had ceased completely and by October of that year the British Joint Services Mission in Washington was in receipt of notification that the US Government intended the reorganisation of weapons and defence programmes and the – albeit temporary – phasing out of the anti-crop warfare programme by 1 January 1958 . In spite of the decision to halt these activities, that year would see the re-activation of the anti-crop warfare programme and the expansion of research and development in this field during the Limited War period (see Table 7.1). This period of activity continued until 1963 when emphasis began to be placed more and more on the military utility of chemical anti-crop agents.

The situation regarding the demise of offensive BW in the UK, and the UK's increasing reliance on the US, was perhaps best summed up by the DRPC in the early 1960s which noted that, 'an offensive capability in BW and CW was essential, for the West, though not necessarily for the UK itself'.[109] This sentiment appears to have applied to both anti-personnel and anti-crop BW weapons. By 1963 the UK was completely reliant on US offensive biological weapons should they have been required in war. However, it will be noted that in conducting a comprehensive programme of anti-crop biological warfare research and development, the US had successfully negotiated the respective steps in acquiring an operational BW capability as identified by the US Office of Technology Assessment (OTA) in Table 1 (Chapter 1, see Stages 1 to 8). This resulted in the assimilation into the US strategic war-fighting arsenal of a number of offensive anti-crop biological weapons systems. It is to the latter that we will now turn.

9
Munitions for Anti-Crop BW Agents

In the open literature, detailed descriptions of BW munitions in general, and in particular, anti-crop BW munitions, are the exception rather than the rule. However, in regard to BW munitions in general a notable contribution to the literature can be found in Volume II of the SIPRI series *The Problem of Chemical and Biological Warfare: CB Weapons Today.*[1] Based on an overview of US biological weapons between 1940 and 1972, this volume contains descriptions of the operating principals of a number of anti-personnel biological weapons systems, and includes references to anti-crop biological weapons systems in the form of spray tanks for use on a variety of aircraft. However, it was not until 1981 that a detailed description of the operating principals of US anti-crop biological weapons systems appeared in the open literature. This notable contribution by Robinson entitled 'Environmental Effects of Chemical and Biological Warfare'[2] discussed the development of such weapons prior to the utilisation of the anti-crop chemical-type spray tanks for use with biological anti-crop agents described above. The paragraphs that follow further elaborate on US anti-crop weapons systems and focus on weapons that saw research, development, and in some cases assimilation[3] into the strategic US war-fighting arsenal, prior to attention being switched to the development of spray tanks.

In the post-war period US biological warfare workers devised a capability to attack an enemy's crops with biological warfare munitions filled with biological anti-crop/plant agents. During the course of work on the development of anti-crop warfare munitions a number of ingenious and extra-ordinary devices saw actual assimilation into the US strategic war-fighting arsenal.

The 20-mm anti-crop BW projectile

One of the munitions that did not result in assimilation into the US stockpile was the 20-mm projectile. Of the items that were to be 'rushed to completion' under a directive[4] issued by the then Chief Chemical Officer General McAuliffe, was the interim 20-mm BW weapon projectile used in the dissemination of cereal rust spores. According to Miller, however, by September 1950, the 20-mm projectile had been withdrawn from development due to 'unsatisfactory results in field trials'.[5]

The M115 (M73R1) anti-crop BW bomb

The military requirement for an anti-crop BW munition was first stated by USAF in a letter to the Chief Chemical Officer dated 26 September 1947.[6] Procurement action was initiated for some 4,800 E73 anti-crop BW cluster munitions in October 1950. This munition was intended to provide the USAF with a capability to attack cereal crops with wheat rust. The E73 anti-crop biological bomb was a modification of a propaganda leaflet bomb designated M16A1.

By 1950 research and development with feathers for the dissemination of anti-crop BW agents was reported to have emerged as a more promising method of getting the agent – causing stem rust of wheat – to the target area than previous investigations into the use of high explosive munitions as a means of dissemination for anti-crop BW agents.[7] According to Miller:[8]

> Development of the feather bomb had made possible an immediate Air Force capability against cereal crops. This bomb utilized the idea of impregnating feathers with BW agents. The feathers acted somewhat like a sponge. Their density was such that they were easily wind-borne, and random dissemination could be easily accomplished. This bomb would employ the agent in dry form (others employed the agent in aqueous suspension). Tests had showed that the M16 leaflet cluster could distribute pathogenic anti-crop agents. The single M16 cluster contained enough particulate matter dusted with the agent to create 100,000 foci of infection within a 50 square-mile area.

Washed, fluffed, white turkey feathers of the following dimensions were used during the course of the tests: 3.5 inches in length; and, 1.75 inches wide. The number of feathers per pound were reported to be 5,300. According to a 'Special Report' published in 1950:[9]

> Such feathers will hold loosely up to 80 percent by weight of rust spores. After violent shaking approximately 10 percent by weight of spores are retained. The high retentive characteristics of the feather are due to its anatomy and to the structure of the spore. The feather contains numerous barbs and hooklets which act as pockets to retain the spores which are likewise barbed. For this test, 1 part spores to 10 parts of feathers by weight were used, but the optimum spore/feather ratio needs to be determined.

Feathers and rust spores were mixed, therefore, to the following ratio: 2.5 lb of feathers; and 0.25 lb of rust spores. The above feather/rust ratio was then placed in a mixing drum and rotated for 30 minutes at 9 (revolutions per minute)

In the early 1950s, US anti-crop BW researchers conducted a series of field tests with the causal agent of cereal rust, *Puccinia graminis avenae*[10], Race 8. The results of the tests conducted at Camp Detrick and St Thomas, Virgin Islands, were published in a *Special Report*[11] marked 'Top Secret' in December 1950. In accordance with the introductory remarks of the report,[12] the purpose of the test was to:

> determine if feathers dusted with cereal rust spores and released from aircraft in an M16A1 cluster adapter (used for leaflets and fragmentation bombs) will permit the transference of spores to cereal plants so that rust infection may ensue.

In the first of a series of three tests, bird cages were distributed throughout a one hundred square foot test plot planted with seedlings of Vicland oats. Prior to being placed in the cages for periods ranging from 1.5 to 24 hours the feathers of live birds had been dusted with the spores of *Puccinia graminis avenae*, Race 8. Upon examination, on all of the plots in question it was reported that heavy rust infection had occurred.

In the second of the tests in this investigation, a test was devised in order to attempt to establish if spores dusted onto the feathers of live birds would be retained in sufficient numbers to cause infection after a flight of 100 miles. After the flight, four of the ten homing pigeons that had been dusted with spores of *Puccinia graminis avenae*, Race 8, were placed in a cage on a 50 square foot plot of Vicland oat seedling for a period of two hours. At intervals during the course of a nineteen day test period, sample feathers removed from two of the four homing pigeons revealed the existence of spores on feathers, with spores remaining viable on feathers up to 19 days. It was further reported that after a 100-mile flight spores were retained in sufficient numbers to inoculate plants and cause primary infection.

The third test was conducted on a series of 4 tests plots at St. Thomas, Virgin Islands. At distances of approximately one 0.5 mile apart, the test plots planted with seedlings of Vicland oats each covered an area of approximately 1600 square feet. In this test, four groups of homing pigeons dusted with *Puccinia graminis avenae*, Race 8, were liberated from aircraft and the birds subsequently returned to their respective plots. After remaining on the plots for periods of approximately two hours, it was reported that heavy infection had occurred on all plots. Spores were therefore retained on the birds in sufficient numbers to cause infection after their release from aircraft.

In a subsequent tests, sufficient quantities of the uredospores of the pathogen, *Puccinia graminis avenae*, Race 8, were collected for testing purposes from infected field plots with the aid of a spore harvester.

A further test was conducted with three feather/agent loaded cluster munitions released approximately one mile upwind of the target area with one of the clusters opening at 1,800 ft above ground and the remaining two cluster opening at 1,300 above ground. It was reported that the feathers were distributed over an area of 7.7 square miles by a wind speed of a reported 15 to 27 miles per hour. Observers reported that feathers were found on only 5 of the 16 plots with 2 plots containing in excess of 100 feathers and the remaining three plots containing only single feathers.

In the period following the tests the extent and severity of infection was monitored for a period of up to 40 days. After a period of two weeks it was observed that cereal rust was present in the eleven plots located nearest what was described as the 'impact area'. Detailed observations continued for a period of up to four weeks after the liberation of the inoculum. With regard to the prevalence and severity of cereal rust the following observations were made:[13]

> Primary infection on the 11 plots attacked initially ranged from 2 to 15 percent prevalence ... with a severity of less than 0.1 percent. ... Three weeks later all plants in these plots were infected by rust and the severity had increased from 7 to 30 percent. Rust continued to increase during the last week of the trial. The 5 plots most distant from the impact area which were not initially infected, developed rust between 1 and 2 weeks following the appearance of rust spores on the other 11 plots. This secondary spread apparently resulted from wind-blown spores.

It was further suggested that actual losses to the crops in the target area amounted to an approximate 30 per cent loss to the grain in the 11 above plots, whereas the remaining 5 plots were reported to have sustained severe injury.

With feathers assigned different colours and loaded into 6 respective cluster adapters in quantities of 10 lb per cluster, a test was devised in order to establish the distribution patterns of the feathers when released from the following altitudes above ground level: 1,300, 1,800, 2,000, 2,300, 3,000, and 4,000 ft. The results of this test are included in the table that follows (Table 9.1):

It was noted that much of the infection observed in the test plot had been caused by spores becoming separated from the feathers. According to the report:[14]

> the 5 plots where feathers were found had infections near carriers as well as some distance away and the remaining 6 plots also developed cereal rust. In those plots where feathers had fallen about 10 percent of the plants were infected with a few scattered pustules. Occasionally 1 plant or more rarely several plants in a group were infected with many pustules. Plants showing many pustules had the type of rust pustule destruction seen when a spore-dusted feather was rubbed on leaves and stems. The scattered infection of a few pustules per infected plant was characteristic of infection following the deposit of free spores.

It was therefore further noted that a relatively heavy primary infection occurring on a plot some 2.5 miles upwind from the impact area had been caused by liberated free spores of the inoculum. The distribution of feathers and liberated *Puccinia graminis avenae*, Race 8 spores was estimated to have covered an area of approximately 25 square miles.

Table 9.1 The results of six drop-tests made to determine the areas of distribution at different altitudes

Opening altitude (approx. feet)	*Average wind speed (mph)*	*Areas of distribution (square miles)*
1300	25	4.6
1800	25	5.0
2000	23	4.6
2300	25	6.5
3000	23	8.2
4000	23	12.5

Further trials with feathers demonstrated that when liberated at an altitude of 1,300 ft feathers were dispersed over an area of 4.6 square miles, and that the range of dispersal increased to 12.5 square miles when feathers were liberated at an altitude of 4,000 ft.

Thus, the report[15] offered the following conclusion:

> It is concluded that feathers dusted with 10 percent by weight of cereal rust spores and released from a modified M16A1 cluster adapter at 1300 to 1800 feet above ground level will carry sufficient numbers of spores to initiate a cereal rust epidemic.

It was estimated that anti-crop BW agent fills for dissemination by the above feather bomb could be produced by the spring of 1951, and the agent fill was standardised by the Chemical Corps Technical Committee on 25 May 1951.

The destructiveness of the causal agent of stem rust of wheat was well known and facilities for the production of this pathogen functioned all year round. According to Miller:[16]

> It was epidemic in nature. It was readily disseminated by the wind, and germination was immediate. The summer, or red, stage was the only stage in the life cycle of stem rust that had any military significance. The spores of stem rust of wheat that were consigned for military use were grown and harvested under carefully controlled conditions. Since the agent lost its viability during storage, new supplies had to be produced during the fall and winter months of each year. Three production sites were operated in the summer; two in winter.

Some 1,800 of the 70,000 munitions in storage underwent modification by the Chemical Corps prior to being suitable for use, with the development work being handled under contract, according to Miller, by the General Mills Company. The M115 was designed for delivery by piloted aircraft.

By 1953 a USAF Air Material Command Plan (12-53) had been devised for the delivery of 3,200 of the above munitions, with the period of operations for the delivery of the munitions limited to: 1 March to 30 May each calendar year.[17] By October 1954 the causal agents of stem rust of wheat and rye were available in refrigerated storage in quantities sufficient to meet Air Force requirements for anti-crop BW. In accordance with Presidential approval, the USAF

Headquarters would issue a directive to the Chemical Corps to transfer agent filled munition components to aircraft for transportation to overseas bases. The agent filled munition components would subsequently be transferred to some 4,800 M115 anti-crop warfare biological bombs that had been pre-positioned at two overseas bases. At no time was the munitions fill stored outside of the United States. According to Miller:[18]

> Therefore, the Strategic Air Command had the capability of conducting biological anti-crop operations as part of a strategic air offensive if directed.

Development of the M115 anti-crop BW cluster munition gave the US its first limited anti-crop biological warfare retaliatory capability.[19] This weapon was shown to be capable of establishing an estimated 100,000 foci of infection in an area of some 50 square miles.

The E77 anti-crop warfare munition

The inspiration for another of the US anti-crop BW munitions in question came from Japan and centred around the deployment of a free-floating un-manned balloon with a biological anti-crop warfare payload. Japanese balloon bombs had been used during the latter stages of the Second World War to carry incendiary and anti-personnel balloon bombs over the Pacific to the North American mainland.

There are but two authoritative accounts[20] of the development of the Japanese munition in the open literature. Mikesh's account of *Japan's WWII Balloon Bomb Attacks on North America,* which was published in 1973 by the Smithsonian Institution was the result of 10 years of painstaking investigation into one of the most obscure munitions to see war-time use. Mikesh's account was followed by a more recent publication in 1992 by Bert Webber, entitled *Silent Siege III.* The connection has not thus far been made, however, between the Japanese munitions and the origins and subsequent development of one of the more unusual weapons to see assimilation into the US arsenal.

The introductory comments to Mikesh's account of Japanese balloon bombs note that this extra-ordinary episode in the history of weapons development was dismissed in some quarters as an imaginative yet pathetically inadequate attempt at retaliation on the part of the Japanese for the daring B-25 air raids on Tokyo conducted by General Doolittle just four months after the Japanese attacks on Pearl Harbor.

But the development and deployment of this munition was of considerable significance. The incendiary bombs caused fire damage, and the loss of six American lives was directly attributable to the deployment of such weapons. The potential psychological impact on war-time America was a cause for concern and reports relating to such attacks were actively suppressed during the course of the War. According to Mikesh[21] this munition was reported to be the 'world's first intercontinental weapon' and the forerunner to its modern equivalent the inter-continental ballistic missile. Never before had the American mainland been attacked with munitions launched from foreign shores. Although the use of this munition was not fully exploited by the Japanese its potential was noted thus:

> Had this balloon weapon been further exploited by using germ or gas bombs, the results could have been disastrous to the American people.

Most notably the US chose to adopt this method of waging war by developing a similar weapon for the dissemination of anti-crop BW agents and assimilating it into the US strategic war-fighting arsenal. According to Franz *et al.*[22]

> Copying the method the Japanese developed during World War II, the Chemical Corps developed the 80-lb anti-plant balloon bomb.

Developments relating to the deployment of Japanese balloon bombs are therefore of relevance to the US anti-crop BW programme.

The US use of balloons in warfare dates back to the Civil War[23] when the US Army's first air arm was created. The Balloon Corps of the Army of the Potomac had used balloons in reconnaissance missions under the command of Thaddeus S. Lowe. While the Union Army disbanded the Balloon Corps in 1863, a Signal Corps balloon squadron was subsequently established in 1887 with balloons used in the observation of Spanish forces, and during the Battle of San Juan Hill to direct artillery fire. It was not until the early 1950s, however, that balloons would again be assigned for a specific military task by the US military.

While the value of balloons as weapons of war was given consideration during investigations by Japan's Military Scientific Laboratory in the early 1930s, large-scale production of balloon weapons in Japan did not begin in earnest until 1942 with research and development assigned to Japan's 9th Military Technical Research Institute.

Early wartime research had demonstrated that balloons measuring some 6 m in diameter could travel a distance of 1000 km from the west to the east coast of Japan flying at an altitude in excess of 26,000 ft for a period of more than 30 hours. A technical problem, which would subsequently be overcome, would restrict the length of time this particular balloon could stay in the air to either day-time or night-time flights only. This problem was related to need to overcome the transition between the expansion of gas and the increase of pressure in the balloon at high altitudes during day-time flights, and the contraction of gas and subsequent decrease in pressure during the course of night-time flights. Research and development with the 6-m balloon, however, had demonstrated the feasibility of long-range balloon flight but this particular program which envisaged the production of some 200 submarine-launched incendiary balloon bombs with a range in excess of 600 miles was phased-out before the technical difficulties associated with long flight times could be overcome.

Prior to the consolidation of Japan's Second World War Navy and Army balloon bomb programmes in 1944, two separate balloon bomb designs had been under development. Navy balloon bomb research and development with a rubberised silk balloon envelope known as the 'internal pressure type' was subsequently discontinued and research was accelerated with the development of a paper skinned balloon design under the Army's 9th Military Technical Research Institute. The Army balloon bomb was identified by its Fu-Go A Type designation.

Investigations showed that during flight daytime temperatures at high altitudes ranged between 0°C and 30°C, with night-time temperatures plummeting to –50° C with a reduction of atmospheric pressure at 30,000 feet decreasing to approximately one-fourth that of atmospheric pressure at sea-level. Increases in daytime temperatures would cause the hydrogen gas inside the side of the balloon facing the sun to rise to the point where the pressure inside the balloon would cause the envelope to rupture. Conversely the reduction of temperature during night-time flight and the loss of atmospheric pressure would cause the balloon to lose altitude and buoyancy.

Further investigations were conducted with larger 10-m diameter hydrogen-filled envelopes fitted with pressure release valves in the base of the balloons. The pressure release valve regulated the pressure inside the balloon as day-time temperatures increased thus preventing the rupture of the balloon at high-altitudes. An ingenious altitude control and ballast release mechanism was devised in order that the balloon would maintain sufficient height at lower altitudes and at lower atmospheric pressures. The altitude control mechanisms would allow the balloon to maintain a height of between approximately 38,000 ft and 30,000 ft during the course of both day-time and night-time flight.

Further technical obstacles would be encountered during the research and development phase of the balloon bombs. Such investigations related to the study of meteorological conditions and patterns of air flow above Japan, and the development of radio reporting equipment for relaying the progress of balloons in-flight back the Japan's balloon bomb researchers.

Meteorological investigations conducted into wind-flow patterns revealed that local wind flow patterns originating from the Asian continent and gaining momentum during the winter months over Japan extended out over the Pacific forming a predictable fast-moving (between 120 and 185 mph) stream of air between Japan and North America. This 'jet-stream' as it became known was strong at levels above 30,000 ft over the south of Japan in October increasing to and maintaining its highest wind-speeds over the Pacific between Japan and North America between the months of November and March. While it was less easy to predict or dictate the direction of a balloons decent at altitudes below 30,000 ft as it approached the North American continent, calculations relating to the course of the balloon and its height and its speed, made it possible to estimate the flight time required for balloons to release their munitions payload directly over the American mainland (see Figures 9.1 and 9.2).

A number of agencies became involved in the development of reliable radio reporting equipment for monitoring the progress of balloons in-flight under stratospheric conditions. Such investigations were conducted in order to determine the direction of travel of the balloon in-flight, its altitude, the temperature of the balloons envelope and the temperature and pressure of gas inside the balloon, and the functioning of the ballast dropping and pressure regulation mechanisms.[24]

Further investigations into the effects of high altitude on the balloons flight mechanisms and into producing balloon envelopes with sufficient durability to withstand the changes in altitude and temperature were also conducted during the course of the research and development phase.

Development and testing with balloons fitted with radiosonde provided sufficient information to allow Japanese engineers to conclude that Japanese balloon bombs launched between the winter months of November and March could cross the Pacific in three days.

Plans for the launching of balloons during the winter months of 1944/45 were formulated on the assumption the some 15,000 balloons

Figure 9.1 Typical balloon flight profile between Japan and North America[25]

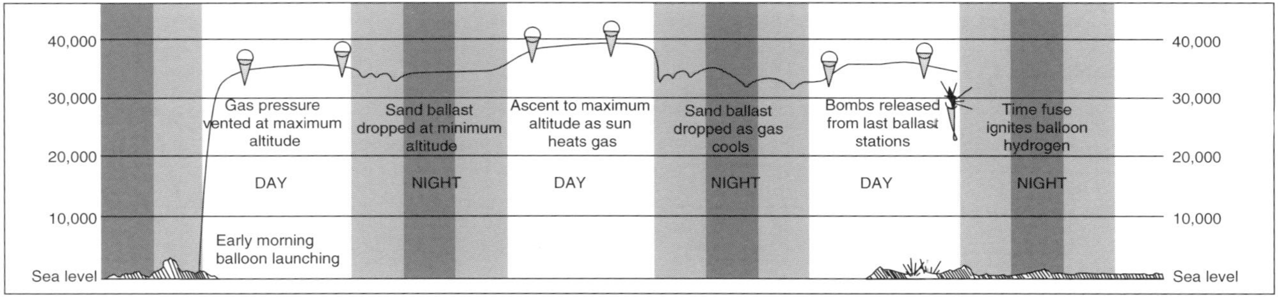

Source: Robert C. Mikesh, *Japan's World War II Balloon Bomb Attacks on North America*, Smithsonian Annals of Flight Series, No. 9, 1973, p. 22, Fig. 25

Figure 9.2 A typical jet stream between Japan and North America during the winter months.[26]

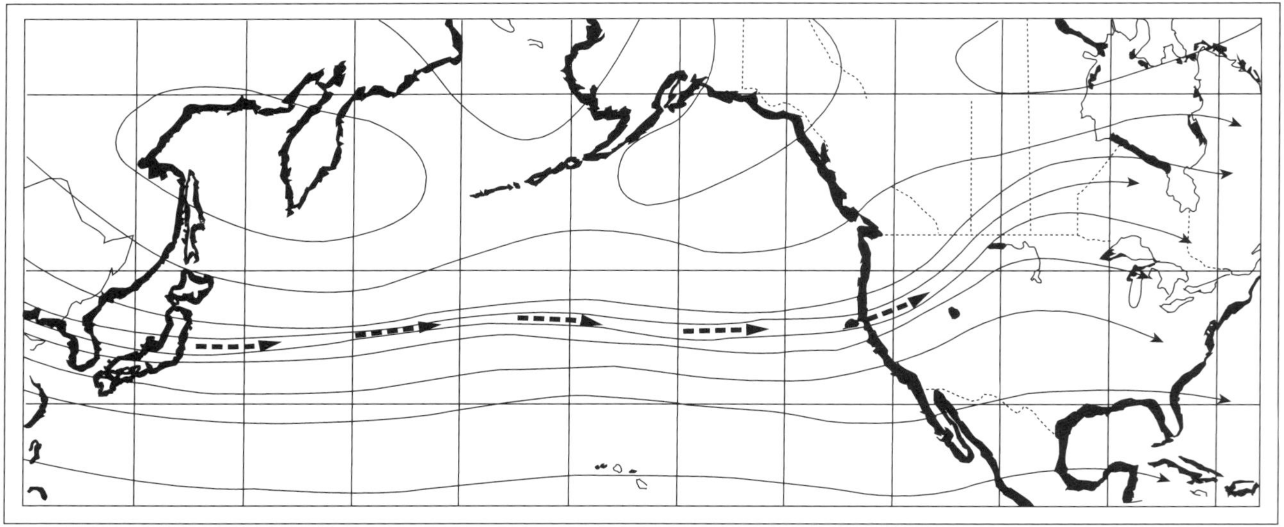

Source: Robert C. Mikesh, *Japan's World War II Balloon Bomb Attacks on North America*, Smithsonian Annals of Flight Series, No. 9, 1973, p. 22, Fig. 25

would be made available. In a programme of accelerated development seven production centres were commissioned in the vicinity of Tokyo and were directed to manufacture the balloons at an estimated experimentation and production cost of some $2 million. Suitable launch sites were identified and a Special Balloon Regiment was established with responsibility for training, the preparation of launch sites with gas production facilities, and for balloon tracking and monitoring. The balloons were launched from three main sites located on the east coast of Japan's southern island. In total some 9,300 Japanese balloon bombs were launched during the winter months of 1944/45 (see Table 9.2).[27]

The 10-m diameter balloon was reported to have had a lifting capacity of approximately 1,000 lb at sea level and approximately 300 lb at 30,000 ft.[28] Balloons could be launched with a munitions payload of a maximum of approximately 70 lb. According to Mikesh:[29]

> Typical maximum loading was one fifteen-kilogram high-explosive bomb or one twelve-kilogram incendiary along with four five-kilogram incendiary bombs.

The munitions payload, which along with sand bags formed part of the balloon ballast, would be released after a pre-determined time once the last sand-bags had been released from the balloon gondola. Release of ordinance would activate another mechanism which ignited a fuse linked to a demolition charge which was designed to bring about the self-destruction of the altitude control/ballast release mechanism. A further flash-bomb charge fixed to the side of the hydrogen filled envelope would then be activated destroying the balloon.

Of the 9,300 balloon bombs launched an undisclosed number of these munitions landed on the North American mainland between the

Table 9.2 Japanese balloon bombs launched between 1944 and 1945

Date	*To be launched*	*Approximate no. launched*
November 1944	500	700
December	3,500	1,200
January 1945	4,500	2,000
February	4,500	2,500
March	2,500	2,500
April (early)	0	400
Total	15,000	9,300

Arctic Circle and the border with Mexico with evidence of the parts of some 285 distinct munitions recovered between November 1944 and July 1945. US military and government officials inspected the component parts of a number of Japanese balloon bombs and one such munition was recovered intact in California and closely examined by the US Army in 1945.

Prior to the completion of a comprehensive investigation into the Japanese balloon bombs a statement was issued by the US War Department in March 1945 enumerating six possible uses for such munitions. Detailed in order of importance, according to the War Department, the balloon bombs might be used to deliver:

1. Bacteriological or CW or both.
2. Transportation of incendiary and antipersonnel bombs.
3. Experiments for unknown purposes.
4. Psychological efforts to inspire terror and diversion of forces.
5. Transportation of agents.
6. Anti-aircraft devices.

Recovery of balloon parts and reports of balloon sighting resulted in preparations for balloon interception by US military aircraft, and balloon intelligence gathering activities under the Western Defense Command with provisions to combat forest and grass fires. In this connection Army personnel were stationed at strategic points in West Coast forest regions in 1945.

In a related contingency plan know as the 'Lightning Project' animal health and agricultural personnel were instructed to monitor livestock and crops for signs of unusual outbreaks of disease and provisions were made for the transfer of decontamination chemicals to strategic locations in the Western regions of the US.

While Japan's balloon bomb raids on North America must be considered a military failure, the potential of such a weapon did not go un-noticed by US BW workers.

However, it was not until 1950 that work began on the US E77 Balloon Bomb. Preliminary military characteristics for a munition based upon the Japanese design had been accepted by the Chemical Corps' Technical Committee in April 1951. According to Miller,[30] 'this munition represented approximately 1/6th of all development effort on biological warfare munitions'. Like the Japanese balloon bomb, the US E77 anti-crop BW balloon bomb consisted of a hydrogen-filled[31] envelope. Suspended from the balloon envelope was a cylindrical balloon gondola measuring 32 inches in diameter and 24 inches in height. The method of impregnating feathers with anti-crop BW agent was also incorporated into the design of the US anti-crop balloon bomb. The balloon gondola, which was designed to house five anti-crop BW containers consisting of a feather and anti-crop agent mix, was grouped around a chemical-type heating mechanism designed to prevent damage to the agent from low temperatures.

In addition to the chemical-type heating mechanism the gondola also contained the following mechanisms: a barometric and mechanical time mechanism designed to eject the munitions payload from the five biological containers at a pre-selected altitude; and, a mechanism designed to neutralise the payload to prevent its dissemination over friendly territory. The characteristics of the fuse mechanism are described in a partially declassified Operations Research Group Study thus:[32]

> The mechanism that controls functioning is a combination barometric fuse and mechanical timer. The mechanical timer controls the drop and neutraliser while the barometric element determines the burst altitude. Drop time may be set for any period up to 60 hours after launching. The desired burst altitude for the falling bomb is preset by means of a calibrated selector switch.

The E77 biological munitions system was designed, according to Miller[33] as a strategic weapons system:

> It was to be launched by a special group assigned or attached to the theatre air commander. Five launching sites were planned. Training was to be the sole responsibility of the 1110 Air Support Group, Headquarters Command, stationed at Lowry Air Force Base, Colorado. A hard core of personnel could be expanded if necessary. Air and Airways Communication Center and Air Weather Service were to cooperate in tracking [the munitions once launched].

The equipment required to launch the balloon system is described in the partially declassified report[34] cited above, as follows:

> Major equipment of the launch site will include a source of hydrogen gas or gas-generating unit and covered wagon launch carts. Prior to inflation the balloon is laced into an Orlon covering on the launch cart. The cover restrains the balloon during inflation and protects it from damage. When the unit is to be launched, the covering is opened down the middle and as it falls away the balloon rises. The entire process can be completed routinely in 30 minutes.

Progress with this munition was hampered from the outset due what Miller identified as 'deficient' military specifications. Additionally, a number of technical problems associated with generators and heating mechanisms were also encountered. The latter for example employed a catalyst that limited the shelf life of this component to a period of some six months. However, engineering problems were overcome.

With regard to the quantity of agent that would be required to achieve a specific area coverage it was found that initial assessment had proved unrealistic and the USAF had found that in order to successfully deliver 150 functioning munitions, some 2,400 munitions would have to be launched during a single season. This requirement dictated that, in order to achieve successful area coverage, some 4,000 munitions would have to distributed among five launching sites. The effectiveness of the E77 was stated in the Operations Research Report[35] thus:

> Under proper weather conditions the agent carried to the ground on the carriers is sufficient to cause high levels of plant infection when impact is on target crops.

Testing of this munition at Vernalis in California where 41 such munitions had been launched between October and December, 1954, demonstrated that the military characteristics for this munition had been met.[36]

The effectiveness of this munition was found subsequently to have been enhanced during testing conducted in 1958 which showed that when distributed in an oil based carrier, rust spores remained viable for longer periods of time where environmental conditions remained unfavourable to the viability of the pathogen.[37] Investigations conducted under contract at the University of Minnesota and studies conducted at Fort Detrick offered the following preliminary conclusions regarding the effectiveness of oils as carriers of cereal rust spores thus:

> a./ Spores carried in oil retain viability on plants longer through periods unfavourable to infection than dry spores;
> b./ They are not as easily washed off plants by rain and dew run-off;
> c./ Germination is not only unimpaired, but is accelerated by some oils, hence, perhaps shortening the required dew period for infection;
> d./ Spores can be stored in suitable oils for periods of at least five weeks with little loss in germinability ...

Further testing with the balloon bomb and the E73 cluster munition revealed that both munitions could effectively disseminate either anti-crop or anti-animal agents, or perhaps most interestingly as Miller had pointed out, 'a mixture of the two'.[38] Although a munition requirement had never been stated for a BW agent/munition system that could be used against livestock, 'considerable work' had been conducted with the causal agent of Hog Cholera. However, more extensive work on anti-animal biological warfare agents had been ruled out due to US government quarantine restrictions which prevented work on pathogens that posed a danger to US livestock.[39] Research on anti-animal agents subsequently took place at the US high-security containment facility at Plum Island.

The US E77 anti-crop BW balloon bomb did not see wartime use and was made obsolete as subsequent developments in munitions design became assimilated into the US anti-crop war-fighting arsenal thus bringing to an end another chapter in the history of balloons in warfare.

The following organisations were involved in the development of this munition: the Air Force Cambridge Research Center (Cambridge, Massachusetts); the Chemical Corps; General Mills Corporation;[40] and, Wright Air Development Center, Wright Patterson Air Force Base.

The E86 anti-crop BW bomb

Subsequent to development and testing with the MII5 a further munition was developed by the US Army Chemical Corps and the USAF for the delivery of anti-crop BW agents. This munition, designated the E86, was based on the same operational principles as the M115 biological bomb but was larger in size weighing in at 750 lb. The E86 munition was intended for delivery from the external carriage of piloted aircraft such as the B-47 and B-52. Development of this munition began in October 1951, when in October 1952 production was assigned under contract to the Ralph M. Parsons Company.[41] In fiscal year 1953, procurement was initiated for some 6,000 E86 cluster munitions, but production of these munitions for operational use was not expected before January 1958. As stated in a partially declassified Operations Research Study Report[42] the characteristics of the E86 cluster adapter and heating mechanism are described thus:

> The cluster adapter is fabricated in two equal sections which open down the long axis of the unit. Inside the sheet steel case is a layer of fibreglass insulation, then a layer of aluminum foil, a 150-watt resistance wire heating blanket with thermostats, and, last, a sheet

> aluminum liner. Spaces are provided for the inclusion of lead ballast. Six agent–carrier containers are nested inside the liner. ... The heating blanket is required to keep the temperatures in the bomb well above freezing while the aircraft is on the way to the target. The blanket contains 150-watt resistance wires and thermostats to keep the temperature between 43 and 75 F. Electric power for the blanket is supplied by the aircraft.

Production of the E86 munition was halted in fiscal year 1953 when the munition requirements were reviewed.

Although information relating to the further development of anti-crop BW munitions is limited it is known that research and development was subsequently conducted into the employment of more effective means of dissemination of anti-crop plant pathogens represented by bulk spray systems such as the Aero 1A and the Aero 2A Airborne Dissemination Units fitted to Navy F3D, and Navy F7U jet aircraft respectively.[43] It will be noted that US developments in the area of munitions, and the associated training and preparation of military personnel dedicated to the deployment of such weapons systems, is consistent with the steps in acquiring an operational BW capability as identified by the US Office of Technology Assessment (OTA) in Table 1 (Chapter 1, see Stages 1 to 8).

10
Targets

In the post-war period, US BW workers conducted theoretical vulnerability assessments of its Cold War adversaries. Two such investigations concerned the former Soviet Union and its European satellites; and Communist China. Data in the following section pertaining to the Soviet Union and its European satellites has been compiled from only partially declassified primary source documentation.[1] Released under the US Freedom of Information Act in 1963 this 1958 document was cut by about 50 per cent during declassification. In particular, material on delivery systems, which might be used for *Puccinia graminis tritici* (for use against wheat), was cut from the original 74-page study. Whereas in the case of China, the technical study[2] upon which the following information is derived was declassified in full. The latter study provides a primary insight into thinking about how best to attack plant crops in China, and therefore also into the particular characteristics that such an attack might have.

Former Soviet Union and Its European satellites

In order to assess the potential vulnerability of the food economy in these countries, it was necessary to consider the following: estimated effects of reductions in dietary intake; the composition of the diet in the countries under consideration; and the effort required to achieve a reduction in a percentage of the national diet in the countries concerned. The information that is summarised below (Table 10.1) estimated the effect on body weight of caloric reductions over given periods of time.

Table 10.1 Body weight % loss for given % reductions in caloric intake over selected time periods*

% Caloric reduction below base level	*Duration of caloric reduction* 3 months	6 months	12 months	18 months
10	5	7	8	9
20	7	11	13	14
30	9	15	18	20
40	12	19	25	28
50	15	25	30	34

Note
*Basic data from Memo on Questionnaire submitted by the Cabinet Committee on World Food Programs, 19 December 1946 by the Food and Nutrition Board, Division of Biology and Agriculture, National Academy of Science, National Research Council.

From the information in Table 10.1 it was, therefore, estimated that a caloric reduction of 50 per cent below the base level for a period of 12 months will result in an estimated 30 per cent reduction in bodyweight.

Table 10.2 estimates the economic, social and political effects of given percentage reductions in bodyweight.*

Although not sufficient in its caloric reduction to be the cause of starvation, it is estimated from the data in Table 10.2 that a 10 per cent reduction in body weight could result in a 10 per cent decrease in the capacity for manual labour, a 10 per cent decrease in capacity for clerical and very light work, moderate discontent, and serious civil disorder and strife. A 20 per cent reduction in body weight is estimated to result in a 100 per cent increase in mortality from all causes, and an estimated 5 per cent of deaths could be accounted for by starvation. The capacity for manual labour and clerical and very light work would be reduced by an average of 30 per cent. It is further estimated that discontent and civil disorder could be very serious. At body weight losses of 30 per cent and above, mortality and death from starvation would increase significantly. The capacity for manual labour and clerical and very light work could be reduced to approximately 10 per cent of its normal level, and while discontent is estimated to be extreme at these percentages of body weight loss, civil disorder would have been reduced to a minimum.

Estimates of average calories per capita per day are provided for the former Soviet Union (USSR), between 1950 and 1957. Meat, potatoes, sugar, vegetables, vegetable oils were estimated to make up 794 calories or 28 per cent of calorific intake per capita per day. The remaining

Table 10.2 Economic, social and political effects of given reductions in bodyweight

	Body Weight Loss (%)					
	5	*10*	*15*	*20*	*30*	*40*
Increase in mortality – % from all causes	*	10	30	100	700	3000
Starvation deaths – as % of population in one year	0	0	0.5	5	20	50
% Decrease in capacity for manual labour	*	10	20	40	90	100
% Decrease in manual labour performance	5	20	40	70	95	100
% Decrease in capacity for clerical and very light work	*	*	10	20	50	90
% Decrease in clerical and very light work performance	*	10	25	50	80	95
Discontent and introversion	Slight	Moderate	Serious	Very serious	Extreme	Total
Social comprehension	*	Reduced	Very reduced	Little	Very little	Nil
Civil disorder and strife	Slight	Serious	Very serious	Very serious	Slight	Nil

Note
*Basic data from Memo on Questionnaire submitted by the Cabinet Committee on World Food Programs, 19 Dec. 1946 by the Food and Nutrition Board, Division of Biology and Agriculture, National Academy of Science, National Research Council.

1944 calories, or 72 per cent of calorific intake per capita per day in the former Soviet Union was estimated to be made up by grain. Average per capita per day calorific consumption was therefore estimated to be 2,693 calories between 1950 and 1957. The report offers the following conclusion: [3]

> Thus, a large fraction of the diet is threatened if wheat can be successfully attacked ... even allowing for large-framed, vigorous-working people in a cold climate, the average daily caloric consumption appears to be higher than that required to maintain the health and labour capability of these people by perhaps as much as 10 percent. Thus, on an 'average' basis, the effect of a given percentage reduction in food availability would be partially absorbed by this excess of consumption over need. This 'cushion' might, of course, disappear under wartime conditions due to the transfer of a substantial fraction of the agricultural labour force to the armed forces and to increased spoilage and waste through disruption and destruction of transportation and storage facilities. If this effect of war on food production can be considered to balance the peacetime 'fat' in the current – and extrapolated – diet, percentage reduction in food supplies through attack of various crops should have direct effects on the populace as indicated [in Tables 10.1 and 10.2 above].

In spite of the absence of information relating to the means of dissemination in this study, we can, nevertheless, follow related thinking on the operational use of both chemical and biological anti-crop agents and the way in which their use was envisaged against China.

The importance of rice and the possible impact of anti-rice warfare against China

A detailed theoretical consideration of a United States anti-rice CW and BW capability against China has been released under the terms of US Freedom of Information legislation. Entitled, 'The Importance of Rice and the Possible Impact of Anti-Rice Warfare', this largely uncensored report numbering of some 185 pages, consists of five sections and subsequent subsections, preceded in reverse order by a Summary, an Abstract, and customary Acknowledgements. According to the Abstract: [4]

> Generally, the objectives of this study are to elaborate on a means of destroying or diminishing selected portions of the rice food supply.

The relevant sections and subsections are discussed in more detail in the paragraphs that follow.

In the eastern countries referred to in the report as the 'Orient,'[5] rice is of considerable significance, both socially and economically. In its unmilled state, it is high in nutritional value and an important source of energy. It is the choice of food for both rich and poor alike and it is estimated that rice is preferred by what is estimated to amount to in excess of half of the population of the world (c. 1958).

In the Orient, tried and tested methods involved in the production of rice persist and the report anticipates that a heavy reliance on manual labour will continue as the most likely mode of its production for the foreseeable future. The economic value of rice stems from the demand for its supply. Rice is both produced and consumed locally and it was estimated that less than five per cent of the world production of rice enters world trade. Of the estimated 66 million tonne of rice produced in China in 1956, it is estimated that less that 0.2 million tonne entered world trade.

The population of China was estimated to be in excess of 500 million, with a rural to urban population ration of 4:1. The majority of the population reside in rural areas where there is estimated to be an average of 1,200 people per square mile, increasing to 2,500 people per square mile in areas of intense agricultural activity. Population pressure on land is estimated to be intense and in comparison to the 3.6 million square miles of land-mass that is estimated to be occupied by China, only some 12 per cent is reported to be suitable for agricultural production, with the production of rice occupying only an estimated 3 per cent of land.

The culture of rice production in two regions in China, in particular, Shanghai and Canton, are such that these regions are estimated to be particularly vulnerable to anti-rice warfare. The target areas under consideration in the report are densely populated, range from 500 to 7,420 square miles, and are estimated to yield between 380,000 and 3,240,000 tonne of rice respectively.

The report estimated that the calorific contribution of rice to the diet is approximately 60 per cent. 1,674 calories per person per day is estimated to be the calorific intake level for one person for one day. It is estimated that a yield of 1 tonne of milled rice would be required to feed 2,090 people for one day. Further estimates regarding reductions in rice yield suggest that an antirice warfare campaign may achieve between an estimated 33 and 80 per cent reductions in calorific intake.

The antirice agents under consideration in the report are: 1. a biologically active chemical, known as butyl 2-chloro-4-fluorophenoxy-acetate; and 2. a naturally occurring fungal plant pathogen known as *Piricularia oryzae*, the cause of the plant disease known as rice 'blast'. According to the Summary of the report:[6]

> These types of agents are considered complementary and not competitive. The chemical agent can be used in climatological or environmental situations unsuitable for the pathogen, however, where the situation is favourable for the disease its use would offer a means of attacking huge concentrations of rice. Amounts of these two types of agent required on target differ considerably and are in favour of the pathogen, e.g., on the order of 225 grams per acre for the chemical versus 0.1 to 0.2 grams per acre for the pathogen. ... It is concluded that, because of its potential, a capability to wage antirice warfare is desirable. It is recommended that such a capability be obtained without delay

The report was compiled and circulated to respective military commands in March 1958, two months after the planned phasing-out of the Department of the Army programme on anti-crop BW research and development. The order to the Chemical Corps to cease funding in this connection came from Lt-General James M Gavin, Chief of Research and Development, Department of the Army, in June 1957.[7] In spite of the planned phasing-out of the anti-crop research programme, it was planned that the Army would maintain its already existing anti-crop BW capability. Prior to this, Chemical Corps research and development into anti-animal warfare had been discontinued in 1954. In the opening remarks of the Introduction to the report, attention is given over to an appraisal of the strategic advantage and deterrent value of maintaining a capability to wage anti-food (both anti-animal and anti-crop) warfare in the face of Communist aggression from both the former Soviet Union, and from China.

The report compares the United States with Asiatic countries and alludes to a number of differences that suggest further the vulnerability of the latter. Whereas in the United States there is an abundance and diversity in the supply of food, and even food surplus, and an abundance of industrialisation, reducing the significance of agriculture as a target in times of war; in Asiatic countries the situation is different with huge populations reliant upon a single staple food crop, little food surplus, a low level of international trade in food, and limited industrialisation as potential targets in times of war. According to the report:[8]

> the importance of being able to attack the food supply, as well as the feasibility and necessity of doing so through direct attack on growing crops, becomes apparent. ... Famine ... provides a real threat; its terrible consequences have been experienced in most Eastern countries, especially China. An ability to induce famine by antifood warfare can provide a strong deterrent to aggression by these countries.

The opening remarks of the Introduction are followed by important caveats with respect to the significance of the data presented throughout the report. For the purposes of this study, limited field-testing had yielded insufficient data for a comprehensive military evaluation of anti-rice warfare, and in certain instances data from other reports had been used in the compilation of data for the study. According to the report, '[t]herefore, estimates of effect are necessarily extrapolations, and are subject to this limitation'.[9] Reference material used in the compilation of data for the report were extracted from a number of sources,[10] and data available from publications in the public domain pertaining to this study are supplemented by data from a number of classified sources.

With regard to the importance of rice to the millions of inhabitants of Asia, Grist is cited thus: [11]

> The fact that rice is the staple food of the great majority of the people of Asia makes it of vital importance to the economy of countries concerned. In the extensive rice-producing areas practically the entire population is ether directly or indirectly dependent on the success of the crop, a dependence that extends beyond the growers of the commercial community, for ability to buy depends on the proceeds of the rice harvest.

Broadly similar conclusions are cited from Efferson:[12]

> Rice is important to everyone regardless of the amount of the food they consume or their interest in the crop. Rice, not wheat, not

> corn, or potatoes, is the most important food crop of the world. Rice supplies the major food requirements for more than one-half of the world's population. This single food product makes up from 70 to 80 percent of the entire food intake in many countries.'

A section of the following paragraph is emphasised in the original documentation. Again according to Efferson:[13]

> *Because of its worldwide use as a food, rice is more important than the atomic bomb in the present period of political unrest. The powers that control the major surplus rice areas are the powers which will control about one-half of the world's population. By the same token, a stable progressive rice industry in the United States is an important diplomatic tool in maintaining world peace and stability.*

Supporting evidence from literature in the public domain is cited in regard to rice varieties thus. From Grist:[14]

> *Oryzae sativa* [rice] is a cultivated genus of *Oryzae* Linn., of the Natural Order Graminae, widely grown in tropical and sub-tropical regions, whether as a dryland crop or more usually in water for the ripe seed, which is the staple food in many Eastern countries. It is not, however, an aquatic plant (in the botanical sense).
>
> Both wild and cultivated forms of *Oryzae* are annuals, although it is said that at least one variety of wild paddy in Ceylon is perennial. The maturation period varies considerably in cultivated forms from two to three months up to nine months.
>
> The term 'paddy' is variously applied to the land on which the plant is grown, to the husked or unhusked seed, to the cooked or uncooked grain.

While varieties of *Oryzae* were reported to number between 7,000 and 12,000 it was reported that most varieties of rice originate from two varieties of *Oryzae sativum*: *indica* and *japonica*.[15]

Consideration was given to the effects of agents under review in the study against an unspecified spectrum of plant species. Against the use of the chemical agents it was reported as follows, that:[16]

> there does not seem to be a real basis for expecting profound differences in susceptibility among the variants of rice today to these

compounds. Any such differences among the multitude of variants would be expected to be relatively minor in most instances. Such differences have not been found to be great among varieties of other species of plants susceptible to these agents ...

With respect to the action of plant pathogens against a variety of species of rice, it was reported that:[17]

In dealing with rice pathogens, some hundreds of rice variants have been examined and results so far definitely have not indicated that a different race of an organism would be required in order to produce a disease on each of the rice variants. Although some differences have been found, it appears at the present that a broad spectrum of rice variants will be found susceptible to a few strains of the organism.

Oryzae sativa var *indica* and *Oryzae sativa* var *japonica,* are reported to be distinct to the following geographical areas with the *indica* varieties cultivated in, 'India and the rice belt of the tropical Asian counties'; and, *japonica* varieties, 'cultivated in Japan, Korea, China, and other rice-growing areas in the sub-tropics.'[18]

Considerable variability is exhibited between the above sub-divisions. The period of maturation between the above is reported to range from between 80 days to approximately 200 days for different varieties. Upland varieties dependent on rainfall having shorter growing periods and lower yields, and lowland, or wetland varieties, grown under irrigation producing greater yields. In relation to the importance of different rice growing areas DeGeus is cited:[19]

World production of wetland rice is much more important than that of upland rice, because the total area under the former is many times greater and the average yields of upland rice are much lower, often only half as much or less.

Additionally Grist[20] is also used to identify the important rice growing areas of China and the climatic characteristics of the regions:

Buck has classified the paddy areas of China, He shows that the rice regions in that country lie between latitudes 23° and 30° North, and extend from sea-level to 1,500 meters, the climate in the rice region is temperate and sub-tropical, considerably tempered by maritime

> influences along the coast. The humidity is high and the rainfall averages about 50 inches. The growing season is about 326 days, the average temperature being 67.1°F. The summers are very hot and humid, but in the winter frosts are experienced in some areas. In the northern half of the area winter crops, such as wheat and barley, are followed by paddy in the summer; in the southern half of the region there are no winter crops, therefore paddy is followed by paddy. Double-cropping extends over two-thirds of the area. Thorp states that is some places a small third crop is obtained in the year by inter-planting before the second crop is harvested. Busk's divisions of the Chinese area are – the double-cropping area of Kwang Tung 113°E. longitude, the South-western area 30°N. latitude, and the Yangtze area 116°E. longitude; south of this is the rice-tea area.

In commenting on aspects of germination that pertain to the growing of rice Grist[21] is cited:

> Properly ripened and harvested paddy should reach a maximum germination of nearly 100 percent when it attains the correct condition of maturity.

Water requirements differ greatly amongst and between the many respective varieties of *Oryzae sativa*. Varieties that are reported to be 'dry' varieties require, however, the guarantee of periodic rainfall over the 4-month period of maturation for cultivation to be possible. Among the varieties reported to be 'wet', cultivation is dependent on a number of factors such as: ability to withstand the movement of water or a preference for stagnant water; and, resistance to flooding and drought. From Grist:[22]

> Graham places such varieties into three main groups in regard to their water requirements, viz. drought-resistant, normal and flood-resistant, and states that the highly drought resistant varieties can exist from twenty days to one month without water, while the longest period that a paddy can withstand flooding is for about fifteen days. He suggests that the probable explanation of the phenomenon that among the late varieties both drought-resistant and flood-resistant varieties occur, may be that the danger of flooding to which the late varieties growing in the lowest fields are exposed in their early stages, and the late date of maturing, occurring long after

> the seasonal rainfall has ceased, have evolved a type which may be both flood-and drought-resistant.

In regard to growth characteristics of different varieties, it is reported that there is no correlation between the number of tillers formed and the number of inflorescence across the range of varieties of rice plants. It is reported, however, that ideally, tillers should ripen at the same time.[23]

Under the sub-section Maturation Period it is reported that this period can range between 90 and 260 days and as a general rule, varieties with a short maturation period are reported to yield smaller crops than varieties with longer maturation periods.

Long standing cultural practices in the production of rice are in evidence throughout the major rice growing regions. Citing Efferson in relation to methods of rice production:[24]

> At least three-fourths of the world's annual rice harvest is accomplished with all ploughing, fitting, seeding, irrigating, weeding, cutting, and threshing done without equipment other than the historic wooden plough and pointed stick.

> the common practice in planting rice is to plough the soil to a minor extent just at the beginning of the rainy season in May. After the next rain, the rice seed is sown by hand over the ploughed land. Then the producer forgets his crop until harvest. The heavy rainy season, starting in June and extending to November, gradually floods the fields and keeps them flooded until the rains stops at about the time the crop is mature.

It is reported that the 'average family'[25] cultivates between 2 and 15 acres of paddy all of which is either consumed or traded locally. Buck (cited by Grist)[26] identifies the most significant rice producing areas of China and the methods employed in the production of rice as follows:

> the larger cultivated areas are in flat irrigable valleys, scattered from Lungchow, Nanning, and Ishan eastwards at an elevation of 500 to 1,000 meters above sea level. The Canton Delta, about 7,000 square miles, is the most important, and is considered the most prosperous in the whole area, and the most intensely cultivated, The area is practically frost-less, the mean temperature being 57° in February, the coldest month, and 84°F. in July, the hottest month. A well-distributed rainfall of about 69 inches is experienced, with rather high humidity.

According to Gourlay (cited by Grist):[27]

> from Manchuria to south of Shanghai only one crop is grown per annum. From this point to Foochow and in central China two consecutive crops are grown. When two crops are obtained by inter-planting, this is achieved by means of the use of one early and one late variety of paddy. Both varieties are sown in the seed bed at the same time, about the middle of April. The early variety is transplanted in the middle of May in rows eighteen inches apart, and a fortnight later the late variety is inter-planted between the rows. The early crop is harvested by the end of July. The second crop, the growth of which has been retarded by the shade of the first, is then able to develop rapidly and is harvested at the end of October.

DeGeus[28] is cited in reference to the production and availability of rice and its significance in relation to the security of the region in the immediate post-war years:

> Following the serious decline during the war years world rice acreage and production have again increased to such an extent that their levels are even somewhat above the pre-war annual average. Nevertheless, the increase in the number of rice-eating people has been so much greater in the meantime that there is still a considerable shortage in rice.
>
> Production of rice should, therefore, be increased, especially by those countries which still depend on imports. This is the more important because on the results of these efforts will depend the maintenance of peace, order and prosperity in most of the areas concerned.

A section on various countries and methods of rice production numbers 20 pages and includes an appreciation of the methods of rice production in the following counties (in the following order): Italy, Spain, Portugal, Greece, Hungary, USSR, France, Taiwan, India, Pakistan, Ceylon, Java, Thailand, Burma, Indo-China, Malaya, Philippines, South America, Egypt, and the USA. It is reported that significant improvement in yields could be materially raised through improvements to irrigation and drainage, improvements in the selection of seed supplies, through the application of fertiliser, and through improvements in methods of cultivation. Citing Grist the report states:[29]

> About 26 percent of the world area under paddy is grown between 0 and 20°, 66 percent between 20 and 30°, and the remainder in latitudes of 30° Consequently it appears at least possible that improved methods might lead in the aggregate to the production from existing paddy areas of a total world crop of 200 million metric tons of rice, or roughly double the present out-turn.

There are reported to be difficulties associated with the compilation of reliable and accurate data with respect to figures for world acreage and production. In this section information in the public domain is supplemented with data supplied by the United States Department of Agriculture, Food and Agricultural Services.

A marginal increase (7 per cent) in world rice production was recorded in 1955-56 with the production of rough rice only slightly in excess of the record harvest previously recorded in 1953/4. This increase was estimated to represent a 20 per cent increase upon estimated world production of rice in the immediate post-war period between 1945 and 1949. World rice acreage in 1956 was estimated to have increased by 6 per cent on the previous year. Approximately 93 per cent of world rice production in 1955/56, however, was attributed to rice production in Asia. Rice production in China is reported to have increased when compared to previous years estimates. Citing Foreign Agricultural Service estimates in regard to rice production in China the report states:[30]

> The record rice harvest of Communist China in 1955–56 is estimated at 145,500 million pounds, a sharp increase above the 1954–55 crop, which was reduced by floods. This production is 10 percent above estimated production in the average prewar (1931–37) period ... [although] rice production for Communist China is still less than the estimated population gain of 14 percent

Estimates for world trade in rice are presented in the report on the basis of United States Foreign Agricultural Service[31] estimates for the period 1955/56:

> It can be seen that only about 4% of the world rice crop enters world trade because of the large proportion of the crop needed for internal consumption. Burma, Thailand, and the United States, in that order, are the largest exporters of rice by far, their production being exceeded substantially by China, India, Japan, Pakistan, and

> Indonesia ... Communist China appears to be a net exporter of rice to the extent of some 429 million pounds to non-communistic counties ... or about 0.3 percent of its estimated production of 132.9 billion pounds for the period [1954–55].

Along side increases in rice production in China for the period 1955/56 modest increases in industrial production are recorded for the same period with an estimated 8 per cent increase on previous years estimates.[32]

The section of world trade[33] is concluded by the remarks that follow:

> Despite the importance of the world trade in rice, the amounts involved are only a very small part of the quantities produced in the major rice growing areas. However, while certain countries are trying to become self-sufficient or surplus-producers, until they have become successful the import of rice are essential. Thus, it does not appear that the blockade of an importer to deny him rice will be very effective by itself regarding total or overall impact on the rice economy of the country, but considered on a smaller or more local scale the effect my well be catastrophic. The blockade of an exporter appears much more likely to disturb the economy of the major Asiatic exporting countries. In either case it would appear to be much more effective to destroy large portions of the growing crop in an enemy country than to resort to a 'rice' blockade.

While it is reported that near consensus exists regarding the nutritive value of rice, data with regard to its consumption is reported to differ considerably.[34] Data from published sources is supplemented in this section by information obtained from intelligence reports.

Intelligence estimates with regard to the availability of rice in China reveal rice shortages and rationing. Food rationing was introduced in towns and cities in August 1955, and between September and November 1955 rules and regulations pertaining to the purchase and distribution of grains were implemented in the rural districts. In towns and cities food was allocated in accordance with residents age and occupation, with the young and those involved in heavy labour occupations receiving the smallest and largest respective allocations.

Low levels of agricultural production caused by adverse weather conditions in the period 1954 to 1955, and inefficient management of Producer's Co-operatives in rural districts is reported to have resulted in farmers in some areas being faced with near starvation.[35] One CIA

Intelligence Report distributed in April 1956 reported that in the Leong Doo region of Kwangtung Province, farmers were required to sell grain to government Co-operatives at a fixed price. It was reported that the grain was then sold back to the farmers in limited quantities at a price that was inflated with a high proportion of produce exported as a source of foreign exchange.

According to the United Nations Food and Agriculture Organisation (FAO), pre-war estimates for average food intakes were 2,234, calories per person per day. Thus citing Shen the report states:[36]

> Obviously, one will find wide divergence from the above average consumption level in such a vast country as China. Nevertheless, this level could serve well to indicate roughly the quantity of food consumption in pre-war China ... [however] the food-consumption level of China since the war has not recovered to the pre-war level of consumption.

Post-war estimates regarding food consumption compiled and distributed by the CIA in March 1956 suggest revised figures for calorific levels based upon a Chinese population of 573.2 million (Table 10.3).

Revised calorific estimates suggest a reduced level of 1674 calories per person per day based on the following rationale: [37]

Table 10.3[38] Revised calorific levels in China, 1931–55

Food	*1931–37 (average) (calories)*	*(%)*	*1953–54 (calories)*	*(%)*	*1954–55 (calories)*	*(%)*
Wheat	338	16.4	273	15.6	305	18.2
Other Grains	491	23.8	402	23	419	25
Rice	731	35.3	618	35.4*	512	30.6
Potatoes	68	3.3	83	4.8	75	4.5
Total Basic Foods	1,268	78.8	1,376	78.8	1,311	78.3
Oilseeds	104	4	100	5.7	89	5.3
Meat, Eggs, Fish**	78	3.8	69	4	70	4.2
Fats and Oils	114	5.5	87	5	90	5.4
Other	143	6.9	113	6.5	114	6.8
Total Quality Foods	439	21.2	369	21.2	363	21.7
Total	2,067	100	1,745	100	1,674	100

Notes
*Raised 0.1 per cent to balance.
**Excludes fat and fat cuts of port, which are listed with fats and oils.

> it is known that, although pre-harvest hunger occurs in Communist China, the population continues to expand. In an absolute physical sense, then, there must be enough food to go around, and, over given periods of time, to take care of greater absolute numbers. On the basis of present nutritional data, however, no one really knows just what figure for the average number of calories per day represents the minimum requirements. The only possible conclusion is that increasing population, demands for industrialization of the economy, and demands for exports of food products increase food requirements at a rate that probably is slightly greater than the rate of increase in production. This conclusion is supported by evidence of the deterioration of the average diet ... and of the apparent increase in the incidence of pre-harvest hunger.

With regard to preference it is reported that tastes differ considerably from rice producing and consuming nation to the next.

Although in its un-milled state rice is high in nutritional value in terms of its carbohydrate, vitamin and mineral content, it is reported that there are improvements to gained from inclusion in the Asiatic diet of other desirable foods. In reviewing the literature pertaining to the nutritional value of rice near universal consensus is reported with a pound of rice containing 1,644 calories of food energy (or 3,625 calories per kg).[39] This figure, however, is thought not to allow for the distinction between the caloric value of rough and milled rice. Based upon data published by the United States National Agricultural Research Bureau relating to the food value of different crop yields in three provinces for the period 1931 to 1937, Shen's calculations were used further to determine a realistic caloric value for a metric ton of rough rice. This resulted in a revised figure of 3,500 calories per kg of rice, and subsequent detailed calculations resulted in the following estimate:[40]

> Allowing for an extraction rate of 70 percent, the caloric value for a metric ton of rough rice [is therefore estimated to be] 2.45 million calories [or 2,450 calories per kg, or 1,111 calories per pound].

The report comments further on the rice-milling process which involves the shelling of rough rice (sometimes referred to as hulling), the removal of the outer bran layers (usually referred to as polishing), and the separation of rice from its by-products (the rice hull is used as fuel and as bran for the feeding of poultry).

Although data on extraction rates had previously been based upon pre-war estimates, a CIA report dated 9 March 1956 had estimated that as a result of grain processing standards that had been introduced by the Chinese Communists in 1950, post-war extraction rates had improved for non-glutinous and glutinous rice from 74 per cent and 70 per cent, to 82 per cent and 78 per cent respectively.[41]

Possible anti-rice agents

In accordance with the Abstract, the objective of Section III was:

> to present information on possible antirice agents ...[42]

It is noted in the introductory comments of Section III that huge tracts of land are given over to the cultivation of single crops and that such concentrations of agricultural production are susceptible to attack from biologically active chemical agents and naturally occurring disease producing organisms (plant pathogens). The possible antirice agents under consideration in the study are: 1. a chemical agent, 4-fluorophenoxyacetic acid, identified by the designation 'KF'; and 2. a plant pathogen, *Piricularia oryzae* Br. and Cav., identified by the designation 'IR'.

The major points of difference between the above agents has been summarised in the following table (Table 10.4).

A general estimate of the effects of such agents on crop production is included in the following paragraph:[43]

Table 10.4 Characteristics of chemical and biological agents

Chemical agents	*Biological agents*
a. Non-living	a. Living
b. Single agent active on several species	b. Single agent active usually on a single species
c. Not spreading in effect	c. Self-propagating or spreading
d. Synthesised industrially	d. Various culture methods
e. Applied usually as a liquid	e. Applied usually as dry spores
f. Rate 45–225 g per acre	f. Rate about 0.1 g per acre
g. Environment not critical for timing attack subsequent disease development	g. Environmental requirements for successful inoculation and
h. Earliest capability against rice	–

In using either type of agent at an appropriate time prior to harvest, the resulting crop damage can be expected to be severe under environmental condition where crop yield would otherwise be high and as these conditions for the growth of the crop were more adverse, the effect of an attack would be somewhat diminished but not necessarily in direct proportion. *The end result, however, in either case would amount to less food for the enemy.*

Chemical characteristics

Agent effectiveness

Research into the operational effectiveness of anti-crop chemicals has revealed that such agents are most active if dispersed as an acid-solvent liquid solution. In the case of early anti-rice chemical agents which too where composed of approximately one-third acid angens and two-thirds solvent (2,4-dichloro- and 2,4,5-trichlorophenoxyacetic acids with tributyl phosphate as solvent – 2,4,-D, and 2,4,5-T respectively), it was reported that such agents would be:[44]

> generally effective against a spectrum of broad-leaved species of plants usually when applied to susceptible species at rates on the order of 0.1 pounds per acre.

Research had shown that certain crop varieties were susceptible to KF compounds, while not susceptible to the early compounds outlined above. At application rates approximating the optimum application rates of the early agents (i.e. 0.1 Ib per acre), certain cereal species were found to be particularly susceptible to KF compounds, with rice reported to be one of the more sensitive crop species.[45]

One interesting characteristic of the KF compounds in regard to its action against rice was reported to be its delayed effects, with plants appearing normal in the immediate post-application period. It was reported that the enemy farmer would be unable to anticipate the failure of a crop in advance, and the catalyst for the failure of the crop would remain difficult to detect.

Optimum times of attack and effects against a single crop of rice with KF or related compounds are estimated to be as follows:[46]

> 1. prior to transplanting from the nurseries, 2. after transplanting, and 3, prior to heading of the crop. For these three times of attack the type of response of the crop to the agent will largely be vegetative in nature for the first two and inductive of sterility for the last.

Limited testing with the above chemical agent and related compounds has revealed that applications of 0.1 Ib per acre have resulted in yield reductions on rice in excess of 90 per cent. The maximum rate of application for KF and related compounds for an effective strike was reported to be 0.5 Ib per acre, or 320 Ib or 0.16 ton per square mile.[47]

The results of field tests published in the Chemical Corps Research and Development Command, Tenth Annual Report on Technical Progress for the 1956 revealed yield reductions of, 'an average of 95 percent' over a period of 12 dates of treatment, when applied during the reproductive stage of development at an application rate of 0.5 lb per acre.[48]

Field testing of the KF compound (4-flourophenoxyacetic acid) had been undertaken with compounds synthesised and produced in only small quantities under laboratory conditions. Under a contract with Olin Mathieson Chemical Corporation the production of the above agent in industrial quantities had been underway. Three stages in this process can be identified, with stage 1 concerning industrial pre-production studies, pilot plant operation and process design, stage 2 concerning plant design and construction, and stage 3 production and stockpiling.

The following costings for this the X-501 Program are estimated as follows:[49]

	Million Dollars
1. Pre-production Studies	1.025
a. Selection of synthesis route	
b. Pilot plant studies	
c. Process design	
2. Plant Design and Construction	5.5
3. Plant Operation and Stockpiling	10.0
Total	16.525

With the first stage of the contract with the manufactures scheduled for completion during the summer of 1957, it was estimated that a quantity in the region of 5,000 lb of agent would be available for further testing by the Chemical Corps. With the successful completion of stages 2 and 3 and a production facility up and running, so to speak, it was estimated that 3,250 tons of agent would be available at a cost of $1.50 per lb, or approximately $3,000 per ton.

Current military capabilities for the dissemination of anti-crop chemical agents pertain to the delivery of agents 2,4-D and 2,4,5-T. The delivery systems are outlined below:

Airforce systems	Delivery aircraft
1. 1000 gallon large	B-29 × 2 tanks
Capacity bomb bay	B-50 × 2 tanks
Spray tank assembly	C-119 × 2 tanks
Pressure 20 lb per square inch	

Navy systems	Delivery aircraft
Navy aero 14B spray tank	F3D × 2 tanks
90 gallon capacity	F7U × 2 tanks
Pressure 100 lb per square inch	

Research conducted at Eglin Air Force Base into the operational capability of a large capacity spraying system for the Air Force B-29 aircraft suggested that maximum utilisation of the agent against broadleaf crop could be achieved at an optimal flow rate of 100 gallons per minute when spayed from altitudes up to 1,000 feet.[50]

The general approach to conducting the above mentioned and subsequent field trials is stated as follows:[51]

> to determine the spray characteristics of the equipment under a variety of conditions using simulant material in some instances, and active agent in others, together with test plans and other sampling devices arrayed in the 'fall out' area, usually along a sampling line. Thus, estimates of area coverage have been obtained but concurrent yield reduction estimates on really extensive plantings have not been obtained.

The following extract offers a rationale for setting the application rate of KF at 0.5 lb per acre:[52]

> Much of the information available from the dissemination trials using chemical anticrop agents is applicable to rates of deposit below that thought at present to be required for the effective use of a KF-related antirice chemical. It does not necessarily follow that information on swath widths, altitudes of release, airspeeds, flow rates, etc., to achieve deposit rates on the order of 0.1 and 0.05 pounds per acre can be extrapolated to give reasonable approximations applicable to deposit rates of about 0.5 pounds per acre. Determinations of the effects of certain variables on particle size will undoubtedly be useful in future trials where the objectives include an investigation of conditions necessary for obtaining sufficiently

high rates of deposit of the antirice agent. For these reasons later estimates of sorties in the section 'Possible Targets' are based only on the amounts of antirice agent required for the land areas indicated at the presently recommended use rate of 0.5 pounds per acre.

Biological characteristics – rice blast

In the section that follows, literature from published sources is supplemented with classified information pertaining to *Piricularia oryzae* as an antirice agent.

Citing Gaumann[53] it is observed that eight conditions are required for an epiphytotic (an epidemic). The conditions are stated as (Table 10.5):

A report compiled by Camp Detrick (Special Report 208) in July 1954 into a severe epiphytotic that had occurred in spring-planted rice in the Everglades region of southern Florida, noted that in terms of its climatic conditions Florida was similar to 'some major Far-Eastern rice-growing areas'.[54]

In this connection a subsequent report submitted to the Eleventh Tripartite Conference[55] in 1956 provided supporting information as follows:

> Field tests conducted in Florida indicate that 70°F. is critical for infection with rice blast. Incomplete information currently available indicates that night temperatures in much of the potential target areas [in China] generally exceed this figure.

Table 10.5 Conditions required for a plant disease epidemic

Environmental factors	*Host plant factors*	*Pathogenic factors*
1. Optimal weather conditions for the pathogen.	2. An accumulation of development of the susceptible individuals.	5. The presence of an aggressive pathogen.
–	3. Heightened disease proneness of the hosts.	6. High reproductive capacity.
–	4. The presence of appropriate alternate hosts.	7. Efficient dispersal.
–	–	8. Unexacting growth requirements of the pathogen.

The following data is taken from a Quarterly Technical Progress Report for the period July, August, September 1956, which involved the inoculation of some 483 varieties of rice with isolates of *Piricularia oryzae* 825 (from Costa Rica), isolate 920 (from the Philippines), and isolate 640 (from Nicaragua).

A summary of results by geographical regions in included below (Table 10.6):

In accordance with the above findings, just under half of the 483 rice varieties were found to susceptible to isolate 640, whereas approximately 75 per cent of the rice varieties were found to be susceptible to isolate 825.[56]

Discussion papers[57] submitted to the Eleventh Tripartite Conference subsequently reported on developments with regard to the selection of rice pathogen isolates as follows:

> Currently 14 distinct races have been identified on the basis of susceptibility of a number of varieties of rice plants. By using mixtures of races it is hoped that virtually all oriental varieties can be successfully attacked by this agent.

Interim Report 130[58] notes the period of greatest susceptibility of rice plants to the pathogen as follows:

> Since the rice plant is susceptible to attack when in the vegetative growth stage prior to heading, it would appear to be more practicable to attack at this stage of growth development. By this method a relatively small amount of inoculum could be used to initiate primary infection. Spores produced from resulting infections would

Table 10.6 Reaction to *Piricularia oryzae* isolates by geographical regions

	640		*825*		*920*	
	*R**	*S***	*R*	*S*	*R*	*S*
Asia	106***	162	5	224	52	212
Europe	2	6	0	8	0	8
Middle East and Africa	9	14	2	21	6	16
Oceana	3	3	2	4	2	4
Caribbean	13	1	1	131	2	12

* Resistant.
**Susceptible.
***Number of Asian rice varieties resistant to isolate 640.

> cause secondary infection in adjacent areas, and by heading time large number of spores would be in the air capable of infecting heads in a large area. Thus, in effect, fields would produce inoculum for their own destruction. ... Insufficient laboratory and field data are in hand to make reliable quantitative estimation for this agent which would justify prediction at this time. However, the potential exists, and employment of level-flight guided missiles offers interesting possibilities for the attack of large rice-growing areas.

Possible targets in China

Section IV of the report, numbers approximately 57 pages, and is entitled 'Possible Targets' in China. The objectives of Section IV, in accordance with the Abstract are:[59]

> to examine the general rice growing areas of China; to present illustrative examples of specific, limited targets in China with associated estimates of effort and effect; and to provide target information useful to scaling operations.

General rice producing areas (A)

Prior to the inclusion of a discussion on possible targets in China, it is reported that a number of factors may determine the success or failure of anti-rice warfare. These factors are listed in the following extract which is taken from Discussion papers submitted to the Eleventh Tripartite Conference in 1956:[60]

> In anti-crop warfare ... it cannot be considered that large acreage of crops which can be attacked with reasonable effort are necessarily profitable targets. In evaluating crop acreage as a target, some of the factors which must be taken into consideration are:
> a. The importance of the particular crop to the enemy economy.
> b. Existence of reserves of this particular crop, and of all other foodstuffs.
> c. Time-lag which will occur before the loss of this particular crop will effect the enemy will to resist.
> d. Susceptibility or resistance of crop to available agents.
> e. Climatology of the potential target area as related to suitability of crop to attack, and timing of such an attack.

f. Distribution of planting, which may or may not favour spread of the disease from field to field when disease-producing agents are employed.'

Rice has been shown to be of particular importance to the economy of China and although Communist China is recorded as being a net exporter of rice, exports are achieved against a backdrop of government controls on the distribution of rice and the implementation of rationing in towns and rural areas. While reliable estimates regarding food reserves in China in the areas under consideration had not been made available in this report, the availability of food reserves, it was argued, would dictate the time-lag before losses to rice crops would begin to impact upon the enemy's will to resist. With regard to the susceptibility of rice to the agents under consideration, limited field testing has demonstrated the debilitating affects of both the biologically active chemical and the pathogen on the production of rice. While post-attack environmental conditions would affect the establishment and spread of the latter it is reported that climatological and environmental conditions for attack with the biologically active chemical agent need not be as exacting. It was noted, however, that prior knowledge of wind direction and knowledge of respective susceptible stages in development of the crop would also be advantageous.

Although it is reported that there is variation as to the exact boundaries of the rice growing regions in the published literature, there appears to general consensus acknowledging the existence of the two major rice-growing subdivisions. In the larger of the two subdivisions, the northern region, one summer crop of rice is grown. In the southern region, it is reported that the prevailing practice is to grow two successive rice crops.

The following areas are considered to be the Chinese rice-producing areas of principal interest to the study: 1. Central Rice – Wheat; 2. Szechwan Basin; 3. South-eastern or Rice – Tea; 4.Southern Double-Cropping; and 5. South-western. Table 10.7 includes estimates of areas planted to rice, and estimates of yields in each of the above respective areas. The following estimates are, however, reported to be illustrative and not definitive (Table 10.7).

Data from published sources is used to illustrate some of the geographical characteristics of agricultural activity in China. Data from Cressey[61] is cited which estimates that 51 per cent of the land area in China is mountainous and hilly and is not suitable for agriculture. Only an the estimated 14 per cent of land in China is attributed to

Table 10.7 Chinese rice areas and production

Area	*Hectares planted to rice (million)*	*Rice production (million tonne)*
Central rice–wheat	8.214	20.553
Eastern Yangtze	2.922	7.294
Anhwei Province	1.122	2.371
Kiangsu Province(including Shanghai)	1.8	4.923
Central Yangtze	5.005	12.723
Marginal	0.287	0.518
Szechwan Basin	2.702	7.69
Southeastern or rice–tea	2.622	6.841
Southern Double-Crop	4.746*	11.775
Kwangtung Province (including Canton Delta)	3.197*	8.182
Kwangsi Province	1.549*	3.593
Southwestern	1.336	2.962
Total (above)	19.62	49.407
Total for China	26.0**	67.0**

Notes
*These figures are expanded in the literature in order to obtain yields per hectare for a given area of land producing two crops of rice per year. Physical land area involved is thus less than the areas shown.
**Data from Foreign Agricultural Service, Circular FR-8-56, 13 November 1956. Preliminary estimate for crop year 1956/57 (Aug–July).

plains areas which are reported to occupy some 425,000 square miles, and it is indicated that less than 15 per cent of China's 3,657,765 square miles is suitable for agriculture.

The following extract gives and indication of the significance of the regions that are subsequently considered in more detail in the report:[62]

> 'The bulk of the Chinese population resides in the plains areas or essentially on agriculturally productive land. Thus, where large cities have developed in agricultural areas, such as Shanghai and Canton, these cities do not add to but rather diminish agricultural production not only by occupying agriculturally useful land but the population contained in them is a tremendous drain on the produce of the surrounding countryside. The neighbouring areas are not able to satisfy the urban food demand and additional food must be imported from either other parts of China or external sources.

For rural areas data cited from Cressey suggests that population densities can be as high as 2,500 people per square mile. This figure suggests an average of 1,200 people per square mile if the cultivated area of China is estimated to be 425,000 square miles and it assumed that the Chinese population is an estimated 500 million.[63]

Further estimates taken from Cressey[64] suggest the rural to urban population ratio and the rural densities of the antirice warfare target areas under consideration:

> The rural to urban population ratio is about 4:1. Three areas have been indicated as having rural densities in excess of 2,000 people per square mile, notably the delta around Shanghai, the Chengtu Plain in Szechwan, and the Canton Delta. Shanghai itself in 1953 presumably had a population of 6,204,417, a residential square mile in the city having 316,160 people.

An estimated 96 per cent of China's population is reported to reside in the south-east of the country which makes up an estimated 36 per cent of China's geographical area.[65]

This section also included information illustrating the significance of the Yangsee river to transportation and communications in China. It was noted that superabundant rainfall had caused flooding and damage to crops on a considerable scale in 1954 resulting in starvation. While bombing techniques for destroying dykes were considered as a possible means of inducing flooding in the Yangsee river basin to bring about the destruction of crops, such methods were subsequently ruled out.[66]

While data relating to some of regions in the above Table 10.7 will be returned to in more detail in a subsequent section, the areas in the vicinity of the following two areas are regarded to be of particular significance as antirice warfare targets in China: Shanghai and Canton. In both areas it is reported that two separate rice crops are grown simultaneously in the same field. This practice is referred to as 'inter-cropping' and is defined as follows: [67]

> Inter-cropping implies two crops to be harvested at different times but grown together in alternate rows in the same field or paddy.

The Canton Delta region is in the province of Kwangtung in the southernmost sub-division, and Shanghai is in the adjoining province of the Kiangsi. The cultural practice employed in the production of rice in the regions in the vicinity of Shanghai and Canton are described as follows:[68]

> In the Shanghai area ... [t]he seedlings of early- and late-maturing varieties of rice are transplanted in alternate rows with the result

> that two crops are brought to maturity in a general region where ordinarily only one crop of rice is successfully matured. Two items stand in contrast to the Canton Delta where a similar inter-cropping of rice is practised: 1. the varieties of rice grown are different, with greater emphasis upon early maturing types, and 2. the reason for the inter-cropping practice is induced by a shortened growing season arising out of a [cool] fall. ... In the amphibious region along the estuarine and ocean margins of the Canton Delta ... there is danger of the rice being injured by salt sea water which may seep into the paddies with the irrigation water in late summer and fall when the flow of fresh water in the river is reduced. To reduce the likelihood of injury from this source two crops of rice are grown by intercultural or inter-cropping methods...The seedlings of an early maturing variety are set out first and in widely spaced rows. This early rice matures in approximately 70 days after transplanting. About two weeks after setting out the early rice, seedlings of a late-maturing variety are planted between the rows of the first. This procedure allows for the transplanting of the first crop a little later and the harvesting of the second crop somewhat earlier than would be true if the two rice crops were grown in succession.

Commenting on the susceptibility of rice to antirice warfare in the above areas it is reported that:[69]

> It appears that a propitious time for attacking both crops simultaneously would be during the first two weeks they are both in the rice paddy.

The region of Shanghai

Data in the section that follows pertains to yield reduction to Chinese rice crops for the biological active chemical agent only. Detailed estimates regarding the plant pathogen *Piricularia oryzae* were not included in this document. The estimated effects of agent KF on rice in the areas under consideration are included in this section in order to illustrate the estimated effectiveness of this form of warfare. Absence of space prevents a rigorous analysis of the calculations upon which the estimates are based, and as such calculations relate to the chemical agent in some considerable detail and do not relate to the pathogen their inclusion is considered to be outside of the purview of this book. Approximations that are of relevance to both agents (such as target areas, and estimated production yields, etc.), however, are included below.

Table 10.8 Estimates of rice production in the vicinity of Shanghai

Total area planted to rice	1.8 million hectares	6950 square miles
Yield in single rice crop area	1.0 tonne per acre	640 tonne per square mile
Yield in inter-cropping rice area	1.5 tonne per acre	960 ton per square mile
Total province rice production	4.92 tonne	

The following data provides estimates for rice production in the region in the vicinity of Shanghai[70] (Table 10.8):

Total province rice production of 4.92 million tonne divided by the total area planted to rice of 6,950 square miles gives the average for combined single and inter-cropping areas planted to rice as 708 tonne of rice to the square mile. From this calculation it was reported that the land area occupied by inter-cropped rice in the vicinity of Shanghai was estimated to be 2,086 physical square miles, with approximately 10 per cent of the total cultivated land producing 28 per cent of rice by inter-cropping.

At the estimated deposition rate of 0.5 Ib per acre for the biologically active chemical antirice agent reductions in crop production in the inter-cropping area are estimated as follows: [71]

> The amount of production destroyed could be expected to be from 0.7 to 1.4 million metric tons of rice or from 14 to 28 percent of rice production ... depending on the fair to optimal timing of the attack, respectively.

The region of Canton

Like the region in the vicinity of Shanghai, the region in the vicinity of Canton is considered to be of particular significance as an antrice warfare target due the cultural practices employed in this region for the production of rice (Table 10.9):

Where it is estimated that the area of inter-cropped rice in the vicinity of Canton is 3,180 square miles, an application rate of 0.5 Ib per

Table 10.9 Estimates of rice production in the vicinity of Canton

Total area planted to rice	3.2 million hectares	10,600 square miles
Yield in successive or double-cropped[72] area	5.837 million tonne	7,420 square miles
Yield in inter-cropping rice area	2.345 million tonne	3.180 square miles
Total province rice production	8.182 million tonne	

acre for an attack with the biological active chemical agent will reportedly achieve a reduction of 1,173 million tonne of rice, or a 14 per cent reduction in province production where it is assumed that the plant susceptibility stages are sub-optimal and 50 per cent destruction is assumed. Where it is assumed that plant susceptibility is optimal achieving complete destruction of the crop, it is reported that 2,345 million tonne will be destroyed equal to an estimated 29 per cent reduction in province production.

The areas given over to the double-cropping of rice in the region in the vicinity of Canton are estimated to be approximately 7,420 square miles. With regard to estimated yield reductions to the early crop, at an application rate of 0.5 Ib per acre for the biologically active chemical agent it is reported that a reduction of 1.3 million tonne will be achieved or approximately 16 per cent of province rice production when plant susceptibility stages are sub-optimal and it is assumed that 50 per cent destruction of the yield will be achieved. Where it is assumed that plant susceptibility stages are optimal and complete destruction of the crop is achieved it is estimated that a reduction of 2.6 million tonne will be achieved or a 32 per cent reduction in province rice production.

Where yield reductions for the late crop are concerned, at an application rate of 0.5 Ib per acre it is estimated that this agent will achieve a 20 per cent reduction to province rice production amounting to an estimated reduction of 1.62 million tonne where plant susceptibility stages are sub-optimal and only 50 per cent destruction of the crop is achieved. Where plant susceptibility stages are assumed to be optimal with complete destruction of the crop achieved, an application rate of 0.5 Ib per acre is estimated to achieve a 40 per cent reduction to province rice production amounting to a reduction of an estimated 3.24 million tonne of rice.

Estimation for other Areas

Table 10.10 includes approximations of rice production for different cultural practices. It is reported that such approximations may be useful for estimating the production of rice in other areas of China.

It is reported that the following calculations[73] can be used to estimate areas of attack where a given caloric reduction per capita per day for a given number of people for a given period of time is indicated.

> In scaling up a theoretical operation, the above tabular entries, if multiplied by the number of calories per metric ton of rough rice

Table 10.10 Approximate rice production for different cultural practices (tonne per square mile)

Degree of land utilization (%)	70	80	90	100
Cultural practice				
Single crop[c]	448	512	576	640
Double Crop Total[d]	717	819	922	1024
Early crop[e]	318	364	410	455
Late crop[f]	399	455	512	569
Inter-cropping total[g]	672	768	864	960
Early crop[e]	298	341	384	426
Late crop[f]	347	427	480	534

Notes
[a]Assuming wet culture for all practices.
[b]70 per cent of these tonnages would represent milled rice equivalents.
[c]At 1 metric ton per acre.
[d]At 1.6 metric tons per acre.
[e]At 4/9 of total.
[f]At 5/9 of total.
[g]At 1.5 metric tons per acre. Proportion for early and late crops taken from information on double-cropping yields.

(about 2.45 million), will indicate rice calories per square mile. ... Suppose, for example, it is desired to determine the size of the attack area in order to achieve the equivalent of reducing the diet of 50 million people by 25 percent for a period of six months (assuming the rice occupies 80 percent of the land in a region where rice is single cropped, that half the rice attacked is destroyed, and that the prevailing total diet is 1,674 c/c/d). Under these condition the equivalent of half the rice in each square mile would be destroyed so that $\frac{512}{2}$ metric tons of rough rice would be lost or (× 2.45 million) the equivalent of 627.2 million calories per square mile. The caloric loss requirement in this instance amounts to 420 c/c/d for 180 days for 50 million people or $3.78 \times 10/^{12}$ calories and after dividing by 627.2 million indicates the area of attack to be 6027 square miles. A reduction by the amount indicated could mean starvation for these people.

Conclusions and recommendation

The following extract relates to conclusion and recommendations included in the report is taken verbatim from the original.

Conclusions

1. The importance of rice as a food is such in the Far East that it is intimately associated with the basic economy of the rice-consuming countries regardless of whether they are net importers or net exporters of this product.
2. The importance rice has as a food allows this product to be commonly used in many instances in lieu of money for the payment of debts, wages, bonuses, taxes, rentals, and the like.
3. Rice has been the preferred and the staple food in the Orient for many centuries and in surviving this test of time is likely to continue in this preferred status for the foreseeable future.
4. The demand for rice is such that in these counties where suitable land, water, and climate are available, if a crop is grown it is probably rice , that is, if it can be grown, it is grown.
5. The culture if rice is such that, for about 90 per cent of its production, water within small dykes is required on the land for a major portion of the growing season, a uniqueness pertaining only to rice of the major foods of these counties and a feature relatively easily identifies from the air.
6. A susceptible period for attacking rice exists shortly after transplanting, an operation readily apparent by aerial observation and suitable as a reference for timing an attack.
7. The intelligent use of an effective antirice agent in war against primarily rice-consuming counties could prove a potent weapon in reducing their capability and will to wage war. The loss of food by such an attack would not be immediate but would actually be felt by the enemy at harvest.
8. Psychologic effects would be produced within weeks if the attack were timed so as to kill the rice plants. These effects would be suitable for exploitation by psychological warfare, offering food in return for capitulation.
9. The earliest antirice capability is believed possible only through the development of a biologically active chemical agent such as currently recommended for the X-501 Program, namely, butyl 2-chloro-4-fluorophenoxyacetate.
10. This chemical agent is complementary to and not competitive with a biological pathogen, rice blast, because it is suitable for use in environmental situations where the pathogen might not be expected to survive and develop adequately, for instance, in the areas where the night temperatures consistently fail to remain above 70°F.

11. Where the environment is suitable for the use of an antirice pathogen is should be used. However, under present circumstances it is problematical when the research phase can be completed with this agent and a capability obtained.
12. [Intentionally omitted.]
13. Mainland China appears to be particularly vulnerable to antirice warfare. The communist regime appears to be making an extreme effort to control the rice produced and to export it where possible to obtain industrial stature even at the expense of the food demands and requirements of the people.
14. An antirice capability is highly desirable and could prove decisive should the next war pose major rice-consuming counties as the enemy and atomic weapons are (1) not used, or (2) are used without quickly achieving enemy capitulation. In keeping with a policy of minimum destruction antirice warfare is practically ideal in that an enemy brought to submission by these means need only be supported foodwise until the end of the next full agricultural season.
15. An antirice capability could provide a much needed deterrent to communist aggression in the Far East.

Recommendation

1. Because of its potential, it is recommended that a capability in antirice warfare be obtained with a minimum of delay.[74]

The above scenario is based upon the assumption that essential prerequisites for a military significant offensive BW capability were in place such that doctrine had been evolved for the use of such weapons, a stockpile of the agents and munitions was readily available, and training and indoctrination in the use of such weapons had been undertaken. Although the above is limited to the development of a theoretical appreciation of the characteristics that an offensive capability in anti-crop warfare might require, it would seem reasonably clear that the above preparations in the 1950s would have allowed the US to conduct large-scale operations in anti-crop warfare. However, in order to inflict meaningful economic damage on an enemy an assault would need to be conducted on a considerable scale. While the study only gives details for the use of the chemical agent, the areas needing to be attacked to achieve significant results are very large. This contention is born out by the estimated number of aircraft sorties (with Navy Aircraft and Aero 14B Spray Tanks) required for the delivery of anti-crop chemical agents onto target crops in the region in the vicinity of Shanghai (approximately 400 sorties required to attack an area of some 2,086 square miles), and by the estimated number of aircraft sorties (again with the above means of delivery) in the region in the vicinity of Canton (estimated to be approximately 1,997 required to attack an area of some 10,600 square miles). Although the amount of agent required on target favored the pathogen (at only 0.1 to 0.2 g per acre) when compared with the chemical anti-crop agent, the area coverage for agents delivered to the above targets would suggest that similar large-scale operations were envisaged for the delivery of anti-crop BW agents.

In conducting anti-crop warfare the US Navy attached particular significance to the advantages enjoyed by carrier based aircraft arguing as follows that:

> In many areas the growing of rice virtually borders the seacoast thus presenting short distances to targets where the adjacent waters are subject to Naval supremacy. A high proportion of flying time could be devoted to spraying the crops. Minimised fuel requirements per sortie and turn-around time would contribute to a most efficient operation.

In addition to the development of bulk dissemination devices for anti-crop biological warfare agents as discussed in the preceding chapter (Chapter 9), a further US Department of the Army publication[75] declassified in November 1969, revealed that a modification (the Model 537 BW Modification Kit) had become available for the Navy Aero 14B Airborne Spray Tank (as discussed in this chapter in relation to the delivery of chemical anti-crop agents) should the delivery of anti-crop biological agents be required.

The discussion will now turn to offer some conclusions as to why offensive anti-crop BW has represented a significant component in BW programmes about which there is publicly-available information.

11
Conclusions

The preceding chapters have shown that investigations into anti-crop BW formed a significant component in BW programmes about which there is publicly available information. French BW activities, curtailed by the German invasion, clearly demonstrate that investigations were conducted in 1939 into BW, which included the consideration of fungal plant pathogens for use against potato crops. In this connection beetles and the use of insects to impair agriculture also appear on the list of French biological warfare 'research under consideration' in the same year. Although offensive German anti-crop BW was officially prohibited by Hitler considerable attention was given to anti-crop BW preparations with research focussing on some 23 plant pathogens and plant parasites. In connection with research conducted in 1944 it also appears that sufficient preparation, included field testing, had taken place with Colorado Beetles that it was not difficult in that year for German BW workers to envisage a German operational capability to conduct offensive anti-crop BW. Regardless of the stated intent of research and development on BW in wartime Germany, and the reiteration of the prohibition on offensive developments in BW from German high command, proponents of offensive anti-crop warfare working in the Agricultural Section of the Wehrmacht Science Division were, according to Deichmann,[1] 'able to combine their war research with their normal activities without attracting attention'. Offensive anti-crop BW research and development indeed took place in wartime Germany but was clearly at odds with, and impeded by, official policy. The potential of plant pathogens and plant parasites in anti-crop BW was also given careful consideration by Japanese BW workers with research focussing on the causal-agents of cereal rusts resulting in the production of such agents on a large-scale. The results of field-testing suggested that large scale destruction of wheat crops could be achieved with small quantities of nematodes acting as agents of destruction. Work on anti-crop BW in the United Kingdom remained of a fundamental nature with research restricted to laboratory investigation toward the attainment of the long-term goal of an effective offensive operational capability in biological warfare. However, between the end of the Second World War and the late 1950s, there emerged an ever increasing dependence on the US for the production of offensive BW agents and munitions. Although no offensive anti-crop biological warfare preparations had been made in the UK over and above those that had taken place in the laboratory, had the need for the use of such agents and munitions emerged, the West, with its reliance on US progress in this field, possessed an offensive capability in conducting this form of warfare. Given the level of scientific and technical expertise as pioneers in the field of BW, it would not be beyond the realm of speculation to argue that the UK could no doubt have chosen to develop such a capability beyond that of research of a fundamental nature should the need have arisen. Additionally, the book has shown that in the late 1980s, Iraq's interest in anti-crop BW had clearly developed to the extent that a quantity of a causal-agent of cereal rust had been stockpiled, and development into an anti-crop BW munition had resulted in a means of dissemination for spreading the agent resulting in the subsequent infection of a large quantity of cereal crop. Further to the case of Iraq, recent concerns over the proliferation of this form of warfare have been raised in connection with offensive BW developments in the former Soviet Union (USSR). Although there is no firm evidence relating the details of anti-crop BW activities in the USSR, recent assertions contained in revelations from a recent defector from the former Soviet Union to the West, suggest that considerable attention was given over to the study of this form of warfare in that country prior to 1992. According to Alibek,[2] a former senior member of the Soviet Union's offensive biological warfare programme, the:

> Russian offensive biological research complex had employed some 10,000 of its 30,000 scientists and technicians on agriculturally related issues.

Furthermore, this contention is borne out in a recent US Government study[3] into combating the proliferation of weapons of mass destruction emphasising the seriousness with which such allegations are being taken. Added to this, official US estimates[4] suggest that currently there may be 10 states in possession of biological weapons.

Whilst this book has shown that offensive anti-crop BW activities had formed a significant component in all known BW programmes, we must now turn to address the question as to why this might be the case? It is the contention of this book that the answer to this question can be found by dividing the history of BW programmes into distinct historical generations – with each generation of programmes defined in relation to the application of increasingly sophisticated scientific knowledge.

It will be noted that the book began with an account of early examples of BW dating back to 300 BC. These examples can be understood to belong to an era where biological sciences had progressed only to a limited extent and the systematic scientific study of disease had not yet been applied to methods of waging war. This period lasted until the end of the nineteenth century when developments in the field of biology began to result in a more sophisticated understanding of the causal relationship between pathogenic micro-organisms and disease. The use of BW which followed what Dando refers to as the 'Golden Age' of bacteriology[5] then belong to a first historical generation in the application of modern BW and include those instances of BW that occurred during the First World War where pathogenic micro-organisms were used in German sabotage operations against both animals and food supplies. According to Wheelis:[6]

> The German programme is of particular interest for several important reasons. It was: (a) the first national programme of offensive biological warfare; (b) the first biological warfare programme of any kind with a scientific foundation; (c) one of only two confirmed instances of an actual programme of wartime use of biological agents ...; and (d) the first and perhaps only large scale trial of clandestine biological attack by secret agents.

Although records detailing French BW activities are incomplete such records show that the proper scientific study of the military potential of BW began in France in the early 1920s.[7] Additionally, this period, whose end coincided with the late inter-war period, and worsening pre-war international relations, saw the opening of a BW research facility in the USSR[8] in 1929, and then later the initiation of a BW programme in Japan in 1934. Although little is known about BW developments in the USSR during this period, BW developments in Japan resulted in the widespread use of this form of warfare in Chinese Manchuria.

A second historical generation of BW programmes began in the 1940s and ended to coincide with the US renunciation of its offensive biological warfare programme in 1969. During this period, and with the benefit of British knowledge of the subject, the US embarked on a major offensive programme of research and development which lasted for a period of some 25 years. In addition to joining forces with the UK on matters relating to research and development of BW, the US also collaborated with Canada. Developments in the field of aerobiology led to a more sophisticated scientific understanding of airborne particles and the behaviour of aerosolised clouds of pathogenic organisms.[9] Such developments featured prominently in regard to the UKs development of an anthrax munition and the subsequent long-term contamination of a Scottish island. Large-scale testing and bulk production of agents was then switched to the US before the end of the Second World War. The US programme included systematic scientific research and development in a wide variety of agents and during the course of the programme a wide range of pathogenic organisms were studied in order to determine their effectiveness against man, animals, and plants.[10] Based upon an analysis of agents in operational use between 1940 and 1983 from the open literature, Geissler[11] was able to show that a defining characteristic of this period was reflected in the extent to which the growth of organised knowledge had been applied to weapons development programmes, which up to 1969 were dominated by research and development into bacteria and fungi as BW agents.

By 1969 the US had assimilated a number of offensive BW weapons into the US war-fighting arsenal for operational use against man and crops.[12] Given such advances, the official US rationale for the renunciation of the offensive programme – that such weapons were of limited military significance – now seems somewhat surprising. According to Meselson,[13] however, in testimony before the US Senate in 1989, the main reasons for such renunciation were as follows:

> first, these weapons could be as great a threat as nuclear weapons; second, they could be simpler and less expensive to develop and produce than nuclear weapons; and, crucially, the US offensive biological weapons programme could be easily duplicated. ... This stark analysis led to the conclusion that our biological weapons program was a substantial threat to our own security.

Such a conclusion led, therefore, to the renunciation of such weapons and the subsequent agreement of the BTWC and its entry into force in 1972.

In this connection, Julian Perry Robinson has commented that Nixon was acting upon the following evaluation:[14]

> that there was no point in putting the huge resources of American technical genius into a technology that could provide poor weak counties with cheap powerful force-multipliers. The fact that biological warfare had this potential was camouflaged – remarkably successfully – behind a depiction of germ weapons as unreliable and militarily useless, a depiction sedulously propagated not least by the negotiators in Geneva.

This brings us into a third historical generation in regard to the application of scientific knowledge to the study of deliberate disease. This coincided with the Soviet decision to expand research and development on offensive BW following the negotiation of the BTWC in 1969, and its subsequent signing as a co-depositary state in 1972. By 1983 the emphasis on bacteria and fungi as BW agents had switched to the study of viruses and of the agents known to be considered as potential BW agents in that year 19 out of a total of 22 agents under consideration were represented by viruses.[15] This shift, according Geissler,[16] could be accounted for as follows:

> many of the viruses in the list of potential BW agents have been previously considered as too dangerous or too difficult to handle for development as potential BW agents. However, by 1983, the advent of genetic engineering techniques ... had changed this situation.

From what is known about offensive BW preparations in the former Soviet Union it is clear that investigations with very dangerous viral agents had occurred to a much greater degree than was possible give the state of knowledge at the time of the mid-century offensive US programme. The Soviet programme had also begun to apply genetic engineering methods, for example, in order to increase the range of antibiotic resistance of *Yersinia pestis* organism, which is the causal agent of plague.[17]

The anti-crop BW programmes of the middle years of this century discussed in this book, therefore, belong to the second historical generation of BW programmes where military developments focussed on bacteria and fungi as offensive BW agents. While research and development in the US, and the subsequent weaponisation and stockpiling of anti-crop BW munitions centred principally around the development of fungal plant pathogens affecting cereal crops, we must speculate with regard to the choice of further work on anti-crop biological agents in the US which may have resulted in weapons systems that used viruses to target crops. Indeed of the anti-crop pathogens under consideration in the US in 1969, Hoja blanca virus, transmitted by the plant hopper, *Sogata orizicola*, was one of the crop agents mentioned in a previously secret *Biological Agents and Munitions Data Book*.[18] In addition to rice, this virus was reported to affect other crops such as wheat, corn, barley, rye, sorghum and various other grasses. (The anti-crop biological agents mentioned in this *Data Book* have been included in Appendix III.) It may also be the case that research with a variety of agents including viruses had taken place in Iraq given the advanced state of knowledge in the biological sciences at the time of the emergence of that programme. Once again, however, we are left to speculate in this connection. One only has to look to the standard texts on plant pathology to see that Iraq's choice of anti-crop BW agent – the causal agent of wheat cover smut – was indeed capable of producing significant reductions in the yields of cereal crops.

The future application of further advances in the biological sciences then raises the question of what might be the characteristics of a new historical generation of BW weapons as the revolution in biotechnology develops and spreads around the world over the next few years?

Commenting upon such developments in relation to the threat posed directly to humans by future BW agents, the US Department of Defence[19] has warned that we might possibly expect to see future BW programmes: where benign micro-organisms have been genetically altered to produce toxic effects; where micro-organisms have been altered to increase their resistant to antibiotics; where micro-organisms have been altered to exhibit greater aerosol and environmental stability; and, where micro-organisms have been altered in order to overcome the immune system. With regard to pathogens that affect plants, it is clear that the plants that produce our staple food crop are being more and more intensively studied[20] at the genome level resulting in a much greater understanding of the vulnerabilities of plants to pathogens. As the revolution in biotechnology and the molecular medicine revolution develop it is clear that a wide variety of other ways in which scientific knowledge could be misused will become available to the weapons designer. Although the development of such munitions remains only a theoretical possibility today, it would be a mistake to assume that such weapons will never emerge and that we can safely afford to ignore their development as a future possibility. At the back of our minds we have to remember that every scientific revolution so far has been used to refine weapons of war[21] and we will have to work hard to prevent the current revolution in biotechnology from suffering the same fate. The question over the possible consequences of the misapplication of modern biotechnology indeed highlight the value of arms control and disarmament measures in this area of potential proliferation. Indeed, if effectively implemented, it is conceivable that a strengthened BTWC could coexist, alongside international prohibitions and controls on pathogens that pose a risk to human health and the environment, as important components of what Pearson[22] has referred to as a 'web of deterrence'. These matters will be addressed in more detail in the chapter that follows.

12 Related International Regulation and Control of Disease against Plants

Given the extent of the US anti-crop BW programme and current concerns over the proliferation of this form of warfare it would therefore seem appropriate to consider how international regimes might be strengthened so as to reduce the risks posed by proliferation. With the possible exception of the BW programme in Iraq and the BW programme in the former Soviet Union, developments in anti-crop BW took place in the countries described in this study prior to the revolution in biotechnology. However, concern surrounding the problem of proliferation must now address the likelihood that pathogens used in illegal present or future anti-crop BW programmes may have been subject to genetic manipulation.

The BTWC and the use of plant pathogens for peaceful purposes

Agreement on the compliance and verification Protocol for the 1972 BTWC, if it occurs, and if the Protocol is not志 watereddown during the course of the negotiation process, will result in the strengthening of the prohibition against BW against crops. It will foster international co-operation and build confidence through transparency through a comprehensive prohibition that will include national legislation establishing the clear norm that development, production, storage, acquisition or use of BW is totally prohibited. The matter of greater scientific and technological co-operation between Parties to the Convention is an important component of the Convention and is covered under Article X[1] of the BTWC. Under this Article, States Parties to the Convention:

> undertake to facilitate, and have the right to participate in, the fullest possible exchange of equipment, materials, and scientific and technological information for the use of bacteriological (biological) agents and toxins for peaceful purposes. Parties to the Convention, in a position to do so shall also co-operate in contributing individually or together with other States or international organisations to the further development and application of scientific discoveries in the field of bacteriology (biology) for the prevention of disease, or for other peaceful purposes.

In 1996, the Final Declaration[2] of the Fourth Review Conference of the BTWC emphasised the importance of the effective implementation of Article X by re-affirming the importance of scientific and technological developments in biotechnology as areas in which cooperation between States Parties might be fostered stating:

> The Conference once more emphasised the increasing importance of the provisions of Article X, especially in the light of recent scientific and technological developments in the field of biotechnology ... which have vastly increased the potential for co-operation between States to help promote economic and social development, and scientific and technological progress, particularly in the developing countries, in conformity with their interests, needs and priorities.

However, a tension exists between the commitment to promoting cooperation between States Parties through the effective implementation of Article X, and Article I of the Convention which prohibits the development, production, stockpiling acquisition or retention of agents and munitions that have 'no justification for protective, prophylactic or peaceful purposes', and; Article III, under which States Parties to the Convention undertake 'not to transfer to any recipient ... and not in any way to assist, encourage or induce any state ... to ... manufacture or otherwise acquire [that which is prohibited] in Article I of the Convention'. On the one hand, the Protocol which seeks to strengthen the effectiveness and improve the implementation of the BTWC, should not be a mechanism which allows, as Rogers et al.[3] argue:

> un-regulated advanced technology that could be used to create weapons of war [to be] transferred from have to have-not states ... [whilst on the other hand] neither should it be used to prevent the legitimate transfer of technology for peaceful purposes.

Additionally, a question mark continues to remain in regard to the exact nature of scientific and technological cooperation and exchange fostered under Article X of the BTWC. States Parties to the Convention have tended to view the implementation of Article X from differing standpoints and differing interpretations on the nature of cooperation and exchange continue to exist. In developing countries, for example, there has been a tendency to view the implementation of Article X under the Protocol in economic terms where the Protocol is seen as representing a mechanism that facilitates the flow of scientific and technological information from the industrialised North to the developing South. Whereas in industrialised countries, there has been a tendency to view the improved implementation of this Article of the BTWC in security terms, in other words, as a mechanism to facilitate co-operation in biosafety and in regard to matters related to public health, not least in improvements in epidemiological surveillance networks concerned with the outbreak of emerging and re-emerging diseases. While a number of matters need to be resolved prior to agreement being reached, concern regarding the spread of biotechnology is being addressed in other international fora. These initiatives open up new areas where regional, national, and international co-operation on matters relating to the control of the products of biotechnology may be possible.

In regard to developments of relevance to plant pathogens, worldwide awareness of the need to control the handling, use, storage and transfer of naturally occurring and genetically modified pathogens that pose a risk to human health and the environment has been increasing. This has been, in part, due to increasing worldwide public-health and environmental concerns. The existence of international mechanisms concerned with the implementation of controls, such as those under the auspices of the Convention on Biodiversity (Biosafety Protocol), a regime which is developing, and the International Plant Protection Convention which is already in force, require that, in regard to the implementation of a future BTWC Protocol, care will need to be taken in order to avoid unnecessary duplication with international activities that are of relevance to the strengthening of the BTWC. Such initiatives are considered in more detail in the paragraphs that follow.

The Convention on Biodiversity (CBD) and the Biosafety Protocol

Efforts to strengthen biosafety through the control of the products of biotechnology have centred around the CBD and the related United Nations Environmental Programme (UNEP) Technical Guidelines on

Safety on Biotechnology. Six years after its entry-into-force there were 175 States Parties to the CBD. However, in 1999 seven States Parties, including the US had not ratified the Convention.[4] The objectives of the CBD are set out in Article I[5] as:

> the conservation of biological diversity, the sustainable use of its components and the fair and equitable sharing of the benefits arising out of the utilization of genetic resources, including by appropriate access to genetic resources and by appropriate transfer of relevant technologies, taking into account all rights over those resources and to technologies, and by appropriate funding.

Under Article II of the Convention[6] 'biotechnology' is defined as:

> any technological application that uses biological systems, living organisms, or derivatives thereof, to make or modify products or processes for specific use.

Since its entry-into-force in 1993 implementation of the CBD has been monitored through a series of Review Conferences. During the course of the First Conference of the Parties to the CBD in 1994 an Open-ended Ad Hoc Group of experts in Biosafety was mandated to consider the 'need for and the modalities of a protocol', and, 'existing knowledge, experience and legislation in the field of biosafety'. At the Second Conference of the Parties of the CBD[7] which was held in Jakarta in 1995 the Conference decided that a mechanism should be developed to address issues related to safety in biotechnology via:

> a negotiation process to develop in the field of safe transfer, handling and use of living modified organisms, a protocol on biosafety, specifically focussing on trans-boundary movements, of any living modified organism resulting from modern biotechnology that may have adverse effect on the conservation and sustainable use of biological diversity, setting out for consideration, in particular, appropriate procedures for advanced informed agreement.

The Cartagena Protocol on Biosafety (Biosafety Protocol) was adopted by the Conference of the Parties to the CBD in January 2000, and a series of interim measures were agreed upon in preparation for its subsequent entry-into-force.

Consisting of some 43 Articles and Annexes the Protocol contains provisions (Article I) relating to the safe transfer, handling and use of

Living Modified Organisms (LMOs). Additionally, Articles 3 to 10 contain important provision relating to the trans-boundary movement of such organisms including provisions for Advanced Informed Agreement.

The provisions set out under the Biosafety Protocol, once implemented, will help strengthen the effectiveness and improve the implementation of the BTWC. Such measures will contribute to the building of confidence, at regional, national, and international levels that novel organisms whatever their origin are being handled, transferred and used safely and for permitted purposes.

The UNEP International Technical Guidelines for Safety in Biotechnology[8] were adopted as a series of interim measures by the Conference of the Parties to the CBD prior to agreement of the Biosafety Protocol. The *Guidelines* comprise some six chapters and seven annexes. The scope of the *Guidelines* is set out in the Introduction which states that:

> The Guidelines address the human health and environmental safety of all types of applications of biotechnology, from research and development to commercialization of biotechnological products containing or consisting of organisms with novel traits.

Chapter II of the *UNEP Guidelines* addresses General Principles and Considerations. Chapter III concerns the Assessment and Management of Risks. Provisions under Chapter IV concern the setting up of national and regional safety mechanisms with measures designed to strengthen, 'national and/or regional authorities/national institutional mechanisms for oversight and/or control of the use of organisms with novel traits'. International safety measures are set out in Chapter V, which emphasises the Supply and Exchange of information stating that countries, 'are encouraged to participate in the exchange of general information about national biosafety mechanisms ... [and that] this form of information exchange can be carried out through direct information exchange, as well as through the creation of an international register or database'.

The provisions set out under the *UNEP Guidelines* are seen as complimentary to the biosafety provisions under the Cartagena Protocol and both sets of provisions are of relevance to the initiative to strengthen the international legal prohibition against BW.

The IPPC and further surveillance and disease reporting mechanisms

While the Biosafety Protocol under the Convention on Biodivesity and the *UNEP Guidelines on Safety in Biotechnology* represent international mechanisms concerned with controlling the impact of living modified organisms on biodiversity, the objective of the IPPC[9] was to control pests affecting plants. The IPPC, which entred-into-force in 1952 was deposited with the United Nations Food and Agriculture Organisation (FAO) in the early 1950s. Its objectives were to:

> secure common and effective action to prevent the spread and introduction of pests of plants and plant products and to promote measures for their control.

The term 'pest' is defined by a 1995 amendment[10] to the IPPC as:

> any species, strain or bio-type of plant, animal or pathogenic agent injurious to plants or plant products.

States Parties to this multi-lateral control regime agree to establish plant protection organisations at the national level concerned with surveillance activities and information-sharing. In particular, States Parties agree to establish a mechanism for reporting the occurrence, outbreak or spread of plant pests to the Commission on Phytosanitary Measures under the United Nations Food and Agriculture Organisation.

Provision for regional cooperation is promoted under the IPPC under Article VIII through the establishment of plant protection organisations in the following regions: Asia and the Pacific; Australia and Pacific Islands; the Caribbean; Central America; European and Mediterranean; North America; Northern South America; and, Southern South America.

Like the provisions set out under the Biosafety Protocol and the UNEP Guidelines such provisions are of relevance to the strengthening of the BTWC.

According to Wheelis:[11]

> the requirements under the IPPC to share information on important plant diseases with other Member States should insure a high degree of transparency with respect to plant diseases of concern to the BTWC.

While the main concern of this discussion is with multi-lateral control initiatives, important developments have also taken place at the national level with a number of initiatives concerned with the development of national measures to promote the safe and permitted use of micro-organisms. According to Pearson:[12]

> It is apparent that in many countries, information is already being collected and submitted to national authorities about activities and facilities handling, using or transferring dangerous pathogens with inspection being carried out to confirm that the regulations are being complied with.

Indeed, national implementation legislation has already been enacted in the European Community, the UK, the US and in other countries.[13]

The multi-lateral control regimes described above are complemented by ad hoc non-governmental activities concerning the surveillance and monitoring of disease outbreaks in crops. One such initiative, the Program for Monitoring Emerging Diseases[14] (ProMEDmail), is an email-based international surveillance and disease reporting mechanism concerning pathogens that affect plants providing extensive and speedy world-wide coverage of disease outbreaks.

Conclusion

The initiatives to develop regimes to safeguard biosafety through the development of mechanisms for improved epidemiological surveillance and co-operation, together with a strengthened international legal prohibition against BW are mutually reinforcing and will help build confidence at the regional, national, and international levels that organisms which pose a risk to human health and the environment are being used safely and for permitted purposes.

However, effective and efficient implementation of these regimes will be required in order that the deliberate use of plant pathogens, whatever their origin, may not go undetected. Indeed, as Pearson[15] has pointed out, '[t]he challenging goal is to identify how these ... international activities can be utilised to contribute to the strengthening of the BTWC'.

The social and economic consequences of a large-scale anti-crop BW attack could indeed be far reaching, however, some comfort can be gained from the insights provided by American efforts to develop usable anti-crop weapons systems which suggest that the characteristics of a significant offensive BW attack would be such that a large-scale attack would be difficult to conceal.

In regard to concern that has been raised recently, particularly in North America, over the vulnerability of crop production to clandestine attacks perpetrated by sub-state actors or terrorist groups, given the widespread practice of restricting crop production to monocultures of a limited number of major food crops, there is a considerable danger that in industrialised countries damage to crops and significant economic losses could be sustained prior to the implementation of measures to prevent the subsequent spread of the disease. However, developing countries that are increasingly reliant on the production of single staple food crops inevitably lack the sophisticated agricultural extension services of the advanced industrialised countries of Europe and North America. Therefore, less developed countries may be particularly vulnerable to this form of warfare.

However, a note of caution must be sounded regarding the possible consequences should current efforts to strengthen the international legal prohibition against BW fail, as Rogers et al.[16] argue, the world could be faced with 'the prospect of lack of control over a major group of weapons of mass destruction during a period of accelerating scientific and technological advancement. The consequence over the next few decades could be the creation of a devastating range of new weaponry, some of which is certain to be aimed at the food crops that feed billions of the world's citizens'.

Appendix I: Wartime Studies into Potential Anti-Crop BW Agents

This Appendix provides an overview of scientific research into the production and control of certain diseases against crops. The US BW research and development programme had made possible experimentation on pathogenic agents on a scale never previously envisaged. Whilst the US BW programme resulted in the attainment of military objectives in the field of BW, the programme also generated a considerable amount of information of relevance to the peaceful use of pathogenic organisms. At this stage of the research and development phase it is only stated intent that distinguishes offensive research from defensive research.[1]

This investigation concerns factors that affect the infectivity of the pathogen which causes brown-spot of rice.

According to Agrios:[2]

> *Helminthosporium* diseases occur throughout the world and are very common and severe on many important crop plants of the grass family (Gramineae) and in some areas, they also cause diseases on apple (black pox) and on pear (blister canker). ... Different species of *Helminthosporium* cause corn leaf blights, brown spot or blight of rice; crown and root rot of wheat, and the leaf spot of wheat; net blotch of barley, stripe disease of barley, and the spot blotch of barley; the Victoria blight and the leaf blotch of oats; eye spot and the brown stripe of sugarcane; and leaf spots or blights of turf grasses, and the crown and root rots (melting out) of turf grasses.

The investigation involved research conducted under laboratory and field conditions at the US BW research and development facility at Camp Detrick, Frederick, Maryland; and at Beaumont, Texas, between April 1944, and October 1945.

The causal agent of brown-spot of rice, *Helminthosporium oryzae* van Breda de Haan affects rice at each respective stage of development. Extensive damage is reported to occur in the seedling stage of development. In the post-seedling stage, further development is impaired by infection to the photosynthetic areas of the plant's leaves, while during the mature stage of plant development, both the quality and the quantity of plant yield is affected adversely. While studies concerning the environmental factors affecting the infectivity of *Helminthosporium oryzae* have been undertaken, it is reported that such studies involved Oriental strains of the pathogen and that further investigation with US varieties are therefore justified.

The investigation into the factors affecting the infectivity of *Helminthosporium oryzae* was organised into the following sub-sections: 1. materials and methods; 2. moisture relations; 3. air temperature; 4. time of day; 5. form of the pathogen; 6. effect of variety of host; 7. effect of age of host on susceptibility; and 8. discussion and summary. A summary of the above will be included in the paragraphs that follow.

Materials and methods

The following varieties of rice were selected for the purposes of the investigation: Onsen and Butte. For experiments in the greenhouse, rice plants were grown in 0.5 gallon plant pots with 15 seedlings per pot. For experiments where it was necessary to grow plants into the mature stage of development, larger (one gallon) pots where used with five plants per pot.

The following inoculum was used during the course of the experiments: a conidial-dust and vegetative mycelium[3] dust. Moist sterilised oats or rye were utilised in the production of inoculum and 3–4 g. of conidia was produced from each respective 300-g dry substrate sample, with 96-7 per cent of the inoculum reported to be capable of germination. The process of producing conidial dust is described in the paragraph that follows: [4]

> Three hundred grams of seeds, placed in a Fernbach flask with 345ml. of water, were steamed for one-half hour, autoclaved for 20 min. at 15lb pressure, allowed to stand over night, and re-autoclaved. When the flasks had cooled, they were inoculated with an aqueous suspension of conidia and incubated at 22–24°C. in the dark for 14 days. Three to five 200-ml. portions of water were added singly to the flasks, vigorously shaken, and the resulting suspension of conidia removed from the substratum by straining through wire gauze into cylinders after each such addition. The cylinders were allowed to stand overnight in a room at 8°C. to allow the conidia to settle. The supernatant liquid was decanted, and the concentrated conidia were collected in a Buchner funnel on coarse filter paper. The filter cake was broken up and dried in a steam oven for 24hr. at 40°C.

A mycelial dust of *Helminthosporium oryzae* was obtained by way the following process:[5]

> Four-litre Pyrex serum storage bottles were used as culture vessels, with three litres of a medium composed of a solution of 10 percent (by volume) blackstrap molasses and 0.5 percent peptone in each bottle. After sterilisation at 15 lb pressure for 20 min. and cooling, the bottles were inoculated with approximately two ml. of an aqueous suspension of conidia. During the incubation period of six days at 27–31°C. the cultures were aerated with sterile air introduced at a rate of 1500 ml./min. The mycelium was collected in a Buchner funnel, the filter cake broken up and dried at 40°C. for 24 hr. The resulting product was ground in a Wiley mill to produce a dust. Both conidia and the ground mycelium were mixed with a filler of limestone (500 mesh) immediately before use.

A series of experiments was subsequently conducted using the above conidial/mycelial mixture, which was applied to rice plants with the aid of an improvised duster. Prior to this experiment the rice plants had been moistened with water dispensed from an atomiser. Plants inoculated in a cabinet, and outdoors, were then placed in a humidity tent where humidity was maintained at high levels.

Three respective methods were adopted for evaluating the degree of infection. The first method took note of the number of lesions appearing per leaf in relation to the known weight of inoculum used in the experiment, and the area exposed. The second method involved the microscopic examination of leaves in order to establish the relative number of conidia and lesions per leaf. The third method evaluated infection against three criteria: light infection; moderate infection; and heavy infection. The latter method was subsequently refined in order that the area of the infected leaves could be compared with a photographic standard that acted as a control.

Moisture relations

A series of experiments were then conducted in order to establish the effect of moisture upon infection, both prior to, and after inoculation. Three of the six plants used in this experiment were placed in a humidity tent for 24 hours prior to inoculation. The remaining three plants were not exposed to humidity. After inoculation both sets of plants were returned to the humidity tent for a period of 24 hours, then left to stand on a greenhouse bench. The experiment revealed, however, no difference in the degree of infection to either the control plants or the plants that had been exposed to humidity prior to inoculation.

In order to establish the effect of water on plants at the time of infection the following experiment was conducted. Three separate sets of plants were used in this experiment. After being placed in a humidifier for a period of one hour, two sets of plants were thoroughly wetted with an atomiser and inoculated with conidial dust. One set was returned to the humidifier together with a third plant which had been inoculated when the leaves of the plant were completely dry. The remaining set, which had been exposed to humidity prior to inoculation, was placed on a greenhouse bench. After a period of three days no infection had occurred to the plant that had been placed on the greenhouse bench, and while infection was reported to be apparent on the plant that was inoculated with dry leaves and placed in the humidifier, infection was reported to be severe on the plant that had been exposed to moisture prior to and after inoculation. In a subsequent duplicate experiment conducted with mycelial dust, similar levels of infection is reported to have occurred. A further experiment estimated that optimum infectivity occurred at a temperature of 22°C after exposure to high humidity for a period of 10 hours.

Two experiments were designed in order to determine the viability of conidia on leaves under greenhouse conditions. In the first experiment, by inoculating 22 pots of rice with conidia and then placing each pot in a humidity tent at intervals of 2 days, it was established that conidia remained viable for 18 days. In a subsequent experiment which extended the time period to 28 days it was reported that conidia remained viable for the period of the life of the leaf. In field experiments, where heavy rainfall occurred immediately after inoculation, it was found that conidia was washed from leaves reducing infection significantly.

Air temperature

An experiment was conducted in order to determine the influence of air temperature upon the severity of infection. Rice plants inoculated with conidial dust were placed in respective water baths at high moisture levels. The temperature of water and air in the water baths was adjusted as follows, to: 15°C, 20°C, 25°C, 30°C, and 35°C. This test, and subsequent comparable tests with mycelium inoculum, suggested that the optimum range for infection was between 20°C and 25°C.

Time of day

In field testing conducted at Beaumont, Texas, an experiment was devised in order to establish the time of day that would result in the greatest amount of infection. Four-feet square test plots planted with 45-day-old Onsen variety of rice in the tillering stage of development were inoculated with a mycelial dust/powdered limestone mixture that was reported to be 40 per cent viable. At the time the plots were inoculated 2–3 inches of water was present in the field. The following table (Table AI.1) shows the severity of infection where the plots had been inoculated on three alternate days. Inoculum was applied to each plot in the mornings, at midday, and in the evenings.

The figures in Table AI.1 reveal that severe infection occurred consistently where plants had been inoculated in the evenings. Where plants had been

Table AI.1 Effect of time of inoculation on the severity of infection of rice by *Helminthosporium oryzae*

Time of inoculation *Day*	*Hour*	*Condition of plants*	*Percentage of leaf covered with lesions*		*Infection*
			Day 6	*Day 8*	
1*	6:30am	Dry	0–1	0–1	Light
	2:30pm	Dry	0–1	0–1	Light
	8:30pm	Wet	0–15	10	Moderate
3**	7:00am	Wet	0–15	10	Moderate
	2:00pm	Dry	0–5	5	Light
	9:00pm	Wet	10–15	25	Severe
5	6:45am	Wet	......	10	Moderate
	2:30pm	Dry	......	0–10	Light
	9:30pm	Wet	......	20–25	Severe

Notes

*Light showers at 3:30 pm and 7:30 pm.

**Light shower at 7:30 pm.

inoculated in the mornings under wet conditions, severity of infection was greater than on the plants inoculated under dry conditions, and generally inoculation occurring under wet conditions produced more infection.

Form of the pathogen

During the course of the experiments it was noted that greater quantities of mycelium of *Helminthosporium oryzae* were required to initiate brown-spot disease of rice. According to the authors:[6]

> it was necessary to apply per unit area a greater weight of the mycelium than of the conidia to produce a given amount of infection. In general, conidia were five to ten times more effective in producing infection than mycelial dust when the two were compared on a weight basis. The infection produced by the mycelial fragments was similar to that produced by the conidia in every respect. The fragments resumed growth on the leaf surface within 24 hr., and the newly formed mycelial tube usually penetrated the leaf epidermis directly rather than through the stomata.

Effect of variety of host

During the course of the investigations an experiment was devised to determine the existence of varietal resistance. Under greenhouse conditions, however, 20 American rice varieties demonstrated no quantifiable difference in resistance to *Helminthosporium oryzae*.

Effect of age of host on susceptibility

A further test was devised in order to determine the susceptibility of rice to *Helminthosporium oryzae* at various stages of development. In order to achieve test plants of different ages rice (var. Early Wataribune) was planted on a weekly basis and then after a period of three weeks on a fortnightly basis until the oldest plants were 85 days old. The plants were then inoculated with dust of conidia, placed in a humidity tent for a period of 24 hours, and then examined after a period of 48 hours on a greenhouse bench. While it was reported that 15-day-old plants exhibited the least infection, uniformity of infection was apparent on the older plants with no significant difference in the severity of infection. Infection reported to be most severe, however, on the horizontal parts of the plant. Comparable results were achieved with Onsen variety of rice. A further test revealed no significant differences in the relative infectivity of mycelium or conidia.

Discussion and summary

The infectivity of *Helminthosporium oryzae* was reported to be greatly influenced by both environmental factors, and the presence of water. Environmental factors were reported to exert their effect upon the pathogen, rather than on the rice plants. Rice plants exposed to humidity prior to inoculation did not exhibit a predisposition to infection. According to the authors:[7]

> This result would be expected, since microscopic examination of leaved indicated that direct penetration of the leaf epidermis occurred more frequently than entrance through the stomata. Factors such as humidity and light which affect stomatal opening would, therefore, not influence the susceptibility of the host to any marked degree. Conditions obtained in a rice field offer relatively uniform moisture and light effects on the host; therefore, the severity of brown-spot disease in any field is dependent upon the suitability of environmental factors to the dissemination of the pathogen and to its infection process.

Experimentation demonstrated that the presence of water influenced infection in the following two ways: 1. by promoting conidial and mycelial adherence to the leaves of the rice plant; and 2. by creating conditions favourable for germination of conidia and mycelium.

While it was noted that brown-spot of rice developed most favourably in seasons where there was a presence of heavy dews, it was reported that rainfall was responsible for washing the inoculum from the leaves of the rice plants.

The initiation of infection occurred after a period of ten hours exposure under conditions of high humidity at 22°C, and the optimum temperature for infection by conidia and mycelium was reported to be between 20°C and 25°C. Field experiments revealed that severe infection resulted from evening inoculations. *Helminthosporium oryzae* was capable of initiating brown-spot disease at all stages of plant development and no varietal resistance to the pathogen was found.

Appendix II

HARRY S TRUMAN
INDEPENDENCE, MISSOURI
July 25, 1969

Dear Congressman Kastenmeier:

In reply to your letter of July 11th, I wish to state categorically that I did not amend any Presidential Order in force regarding biological weapons nor did I at any time give my approval to its use.

With all good wishes,

Sincerely yours,

Honorable Robert W.Kastenmeier
House of Representatives
Washington, D. C. 20515

Note
President Truman's signature appeared in the original letter.

Appendix III

Table AIII.1 Fort Detrick pertinent technical data on anti-crop biological agents for FY 1970*

	Agents in Category A	*Agents in Category B*	*Agents in Category B*
Crop disease	Stem rust of wheat	Rice blast	Stripe rust of wheat
Causative agent	*Puccinia graminis* var. *tritici* Erikss. & Henn,. race 56	*Piricularia oryzae* Cavara, races 11 and 25 (other races depending upon target area)	*Puccinia striiformis* West
Hosts	wheat, barberry, and certain grasses	rice, possibly some other grasses to limited extent	wheat, barley, various grasses
Infection particles	spores	spores	spores
Dissemination	airborne	airborne	airborne
Particle size	ellipsoid, 18–20 by 22–28 microns	5 by 12 microns when dry	approximately 22 microns average diameter
Bulk density	0.5 g/c.c.	average 0.20, range 0.18 to 0.25 g/c.c.	0.5 g/c.c.
Storage life	approximately 3 years at 4°C	over 10 years at 4°C	two years a 4°C, indefinitely at cryogenic temperatures
Incubation period	6 to 14 days	4 to 7 days	12 to 14 days
Infection dose	0.1 g/acre (1 lb per 10 square miles)	0.1 g/acre (1 lb per 10 square miles	0.1 g/acre (1 lb per 10 square miles)
Viability in aerosol	spores can remain viable in an aerosol for several days	viable for days	several days
Production method	spores are produced on living wheat plants	deep-vat mycelial production followed by thin-layer sporulation and drying	spores are produced on living wheat plants
Military specification	MIL-S-51036B		
Current cost	$50 to $75 per lb ($110 to $165 per kg in an established production operation		

Table AIII.1 Fort Detrick pertinent technical data on anti-crop biological agents for FY 1970* (*contd.*)

	Agents in Category C	*Agents in Category C*	*Agents in Category C*
Crop disease	Hoja blanca of rice	Bacterial leaf blight of rice	Downy mildew of poppy
Causative agent	Hoja blanca virus	*Xanthomanas oryzae* Uyeda and Ishiyama	*Peronospora aborescens*
Hosts	rice, wheat, corn, barley, rye, sorghum, and various other grasses	rice and various grasses	species of *Papaver* and *Argemone*
Infection particles	virus particles in insect vector	bacterial cells	spores
Dissemination	transmitted by a plant hopper, *Sogata orizicola*	seeds, soil, wind, water	airborne
Particle size	n/a	cells 0.3–0.45 by 0.7–2.4 microns	14–8 by 18–25 microns
Bulk density	n/a	not known	not known
Storage life	maintained through viruliferous insects idefinately	two years at 4°C in dried exudate	to be determined
Incubation period	n/a	10 to 28 days	less than seven days under optimum conditions
Infection dose	n/a	not known	not known
Viability in aerosol	n/a	not determined, probably several days	not known
Production method	production of insects on host plants	deep-vet production postulated	obligate parasites, production on host crop
Military specification			
Current cost			

Note
**Biological Agent's and Munitions Data Book*, Miscellaneous Publication 34, Systems Analysis Division, Department of the Army, Fort Detrick, Maryland. November 1969, ch. 11, p. 333.

Notes

1 Introduction

1. SIPRI Yearbook, *World Armaments and Disarmament*, Oxford University Press, 1989, p.100.
2. United States, *Technical Aspects of Biological Weapon Proliferation*, Office of Technology Assessment (OTA), 1993. The OTA was a non-partisan analytical agency that assisted Congress with complex technical issues. Funding for the OTA was withdrawn by the 104th Congress in 1995.
3. In addition to the OTA's stages in the acquisition of a militarily significant biological warfare capability reproduced here as a Table 1.1, other commentators have also identified respective stages in this process. For example in 1996, Meselson submitted evidence to the US National Academy of Sciences outlining some 27 respective stages. See J.P. Perry Robinson, 'Some Political Aspects of the Control of Biological Weapons', *Science in Parliament*, Vol., 53, No. 3, May/June, 1996, p.10. The number of stages in Meselson's list had increased to 28 when the list was reproduced again in 1997 – see M. Meselson, 'Background Notes on Biological Weapons, Department of Molecular and Cellular Biology, Harvard University, 20 August 1997, p.13. Similarly, a list entitled, 'Stages in the Development of an Offensive Capability', adapted from a list published in SIPRI, *The Problem of Chemical and Biological Warfare, Volume V: The Prevention of CBW*, 1971, was reproduced by C. Piller and K. Yamamoto, with some 24 respective stages. See, The US Biological Defense Research Program in the 1980 's: *A Critique*, in S. Wright (ed.), *Preventing a Biological Arms Race*, MIT Press, 1990, p.142. However, for the purpose of this discussion, the OTA's simplified version of the stages in the acquisition of a military significant biological warfare capability has been adopted.
4. J.A. Poupard and L.A. Miller, 'History of Biological Warfare: Catapults to Capsomers', *Annals of the New York Academy of Sciences*, 1992, 666, pp.9–19.
5. Ibid., pp.9–19.
6. Ibid.
7. M. Wheelis, 'Biological Warfare Before 1914: The Prescientific Era', in E.Geissler and J.E. van Courtland Moon (eds), *Biological and Toxin Weapons Research, Development and Use from the Middle Ages to 1945: A Critical Comparative Analysis*, SIPRI Study No.18, Oxford University Press, 1999.
8. J.A. Poupard and L.A. Miller, op. cit., p.13.
9. M. Wheelis, op. cit.,
10. Ibid.
11. M.R. Dando, 'Technological Change and Future Biological Warfare', paper presented at a conference on 'Biological Warfare and Disarmament: Problems, Perspectives, Possible Solutions', UNIDIR, Geneva, 5–8 July 1998.
12. Lepick, O., 'French Activities Related to Biological Warfare: 1919–1945', in E. Geissler and J.E. van Courtland Moon (eds) , *Biological and Toxin Weapons Research, Development and Use from the Middle Ages to 1945: A Critical Comparative Analysis*, Oxford University Press, SIPRI Study No.18, 1999.

13. See: *The Problem of Chemical and Biological Warfare: The Rise of CB Weapons*, Vol. 1, 1971; *The Problem of Chemical and Biological Warfare: CB Weapons Today*, Vol. II, 1973; *The Problem of Chemical and Biological Warfare: CBW and the Law of War*, Vol. III, 1973; *The Problem of Chemical and Biological Warfare: CB Disarmament Negotiations*: 1920–1970, Vol. IV, 1971; *The Problem of Chemical and Biological Warfare: The Prevention of CBW*, Vol. V, 1971; and, *The Problem of Chemical and Biological Warfare: Technical Aspects of Early Warning and Verification*, Vol. VI, 1973.
14. J. Tucker, 'Strengthening the BWC: Moving Toward a Compliance Protocol', *Arms Control Today*, Jan./Feb., 1998, p.20.
15. Theodore Rosebury, *Peace or Pestilence*, McGraw-Hill 1949, p.55.
16. R.M. Page, E.C. Tullis, and T.L. Morgan, 'Studies on Factors Affecting the Infectivity of *Helminthosporium oryzae*', *Phytopathology*, 37: 281, May 1947.
17. Tullis, E.C. *et al.*, *The Importance of Rice and the Possible Impact of Anti-rice Warfare*, Technical Study No. 5, Office of the Deputy Commander for Scientific Activities, Biological Warfare Laboratories, Fort Detrick, Maryland, March 1958, 58-FDS-302.

2 Iraq and UNSCOM

1. United Nations Security Council Resolution 687, S/RES/687/8 April 1991, adopted by the Security Council at its 2,981st meeting, on 3 April 1991.
2. Ibid.
3. United Nations Security Council Resolutions: 660, 661, 662, 664, 665, 666, 667, 669, 670, 674, 677, 678, and 686.
4. United Nations Security Council Resolution 687, op. cit., paragraph 9(a) and (b)(i).
5. United Nations Security Council Resolution 707, S/RES/707, 15 August 1991, adopted by the Security Council at its 3,004th meeting, on 15 August 1991.
6. Ibid., paragraph 3.
7. Ibid., paragraph 3 (v).
8. Plan for Future Ongoing Monitoring and Verification of Iraq's Compliance with Relevant Parts of Section C of Security Council Resolution 687 (1991), S/22871/Rev.1, 2 October 1991.
9. United Nations Security Council Resolution 715, S/RES/715, 11 October 1991, adopted by the Security Council at its 3,012th meeting, on 11 October 1991.
10. Ibid., para. 5.
11. The public record regarding Iraq's programme of weapons of mass destruction contains only selected information from Iraq's Full Final and Compete Disclosures (FFCDs) to the Security Council. A former United Nations Official explains the rational behind the reluctance on the part of UNSCOM to include all of the Special Commission's finding in the public record as follows: 'the public record could not contain all that UNSCOM had done or knew about. If it had, there would have been no pressure on Iraq to come clean; it would merely have adapted its fabricated account of the past to what it knew (through the published record) UNSCOM knew. Therefore UNSCOM had to keep a large part of its activities and knowledge out of the

public record'. Tim Trevan, formerly Special Advisor the Executive Chairman and Spokesman for the UN Special Commission, January 1992–September 1995, letter to Professor Paul F. Rogers, 10 November 1997.
12. UNSCOM, *United Nations*, S/1995/864, 11 October 1995.
13. Ibid., para. 42.
14. G.S. Pearson, *The UNSCOM Saga: Lessons for Countering the Proliferation of Chemical and Biological Weapons*, May 1997, ch. 4, p.7.
15. UNSCOM, *United Nations*, S/1995/864, op. cit., para. 57.
16. It is thought that 'choline' was the name given by Iraq's BW workers for an intermediate.
17. Ibid., para. 59.
18. Ibid., para. 60.
19. Ibid., para., 75(J).
20. Ibid., para. 75 (l).
21. Ibid., paragraph, 75(M).
22. Ibid., paragraph 75 (n).
23. The utility of munitions loaded by Iraq with BW agents remains unclear.
24. Ibid., para. 75 (w)–(x).
25. Ibid., para. 28. According to UNSCOM, '[a]uthority to launch biological and chemical warheads was pre-delegated in the event that Baghdad was hit by nuclear weapons during the Gulf War. This pre-delegation does not exclude the alternative use of such a capability and therefore does not constitute proof of only intentions concerning second use'.
26. Zilinskas, R.A. 'Iraq's Biological Weapons: The Past as Future', *Journal of the American Medical Association*, Vol. 278, No.5, Aug. 6, 1997.
27. UNSCOM, *United Nations*, S/1995/864, op. cit., para. 76.
28. Kadlec, Lt Col. R. p.'Biological Weapons for Waging Economic Warfare, in Battlefield of the Future: 21st Century Warfare Issues', <http://www.cdsar.af.mil/battle/bftoc.html>.
29. UNSCOM, *United Nations*, S/1995/864, op. cit., para. 75 (0).
30. Letter from Steve Black, Historian, UNSCOM, 28 November 1995.
31. Peter Beaumont, Gerald H. Blake and J. Malcolm Wagstaff (eds), The Middle East: A Geographical Study, 2nd edn, David Fulton, London, 1988, p.482.

3 The Study of Disease in Plants

1. E.C. Stakman and J.G.Harrar, *Principles of Plant Pathology*, The Ronald Press, New York, 1957, p.12.
2. Ibid., p.13.
3. George N. Agrios, *Plant Pathology*, 3rd edn, Academic Press, Inc. New York, 1988, p.9.
4. E.C. Stakman and J.G. Harrar, op. cit., p.35.
5. Ibid.
6. According to Stakman and Harrar, 'A plant pathogen is a living organism that produces disease in plants. By extension of the definition viruses are also called pathogens, although they are not usually considered as living organisms', ibid., p.67.
7. George N. Agrios, op. cit., p.6.

8. Ibid., p.7.
9. Ibid., p.6.
10. E.C. Stakman and J.G. Harrar, op. cit., p.38.
11. Ibid.
12. David W. Parry, *Plant Pathology in Agriculture*, Cambridge University Press, 1989, p.12.
13. A parasite invade the host plant on which it feeds and within which it proliferates. See George N. Agrios, op. cit., p.41.
14. Like parasites, saprophytes invade, feed, and live on host plants but saphrophytes can live on either living or dead organic matter, ibid.
15. Ibid., p.269.
16. J.E. van der Plank, *Plant Diseases: Epidemics and Control*, Academic Press, New York, 1963, p.212.
17. George N. Agrios, op. cit., p.275.
18. Ibid., p.511.
19. Ibid., p.519.
20. Ibid., p.516.
21. Ibid., p.622.
22. Ibid., pp.636–7.
23. Ibid., p.640.
24. Ibid., p.645.
25. Ibid., p.703.
26. Ibid., p.710.
27. Ibid., p.197.
28. Ibid., p.645.
29. E.C. Stakman and J.G. Harrar, op. cit., p.344.
30. Ibid.
31. S.R. Chapman and L.P. Carter, *Crop Production: Principles and Practices*, W.H. Freeman, San Francisco, 1975, p.281.
32. Ibid., p.260.
33. Ibid., p.432.
34. E.C. Stakman and J.G. Harrar, op. cit., p.362.
35. Ibid., p.366.
36. It must be noted that a number of qualifications must be made in relation to the reliability of crop loss estimates. A number of difficulties are associated with the measurement of pests (including plant pathogens, animal pests and weeds), crop yields, and the interaction between pests and plants. Some of the difficulties association with measuring crop losses are listed by Oerke *et al.*, as follows: 'The distribution of pests in space and time, the response of plants and pests to different climates and soils, interactions between animal pests, pathogens and weeds are only a few of the complications in crop loss assessment, not to mention the economic and social factors involved. ... On a global level, the relative paucity of information available, the irregular distribution and the varying quantity of data add to the problems in compiling loss figures for the major cropping areas.' See E.C. Oerke, H.W. Dehne, F. Schonbeck and A. Weber, *Crop Production and Crop Protection: Estimated Losses in Major Food and Cash Crops*, Elsevier, Oxford, 1994, p.72. Hereinafter referred to as Oerke *et al.*
37. George N. Agrios, op. cit., p.23.

38. Ibid., pp.24–5.
39. Ibid., p.25.
40. Ibid.
41. E.C. Oerke et al., op. cit., pp.750–1.
42. Ibid., p.751
43. George N. Agrios, op. cit., p.23.
44. Oerke et al., op. cit., p.762.

4 Anti-crop BW and the BTWC

1. According to John van Courtland Moon, by the end of this period, the precedent against the use of poison in warfare had rendered the prohibition a principal of international customary law. See J. van Courtland Moon, 'Controlling Chemical and Biological Weapons Through World War III, in R.D Burns (ed.), *Encyclopaedia of Arms Control and Disarmament*, Vol. II, Charles Schribner's Sons, New York, 1993, pp.567–674.
2. A set of general precepts of the laws of war. General rules incorporated into the customary law of war. See Introduction, SIPRI, *The Problem of Chemical and Biological Warfare: CBW and the Law of War*, Vol. III, Humanities Press, New York, 1973.
3. M. Dando, 'The Development of International Legal Constraints on Biological Warfare in the 20th Century', *The Finnish Yearbook of International Law*, Kluwer International, 1999, p.10.
4. 'Geneva Protocol for the Prohibition of the Use in War of Asphyxiating, Poisonous or other Gases, and of Bacteriological Methods of Warfare'. Signed at Geneva on 19 June 1925.
5. M. Dando, *Biotechnology, Weapons and Humanity*, Harwood Academic Publishers, (for the British Medical Association), 1999, p.14.
6. Ibid., p.15.
7. A. Boserup, *The Problem of Chemical and Biological Warfare: Volume III: CBW and the Law of War*, SIPRI, 1973, p.25.
8. E. Geissler, J. E. van Courtland Moon and Graham S. Pearson, 'Lessons from the History of Biological and Toxin Warfare', in E. Geissler and J.E. van Courtland Mood (eds), *Biological and Toxin Weapons: Research, Development and Use from the Middle Ages to 1945*, SIPRI Chemical and Biological Warfare Studies No. 18, Oxford University Press, 1999, p.256.
9. US ratification of the Geneva Protocol included the following reservation. 'The protocol shall cease to binding on the government of the United States with respect to the use in war of asphyxiating, poisonous or other gases, and all analogous liquids, materials, or devices, in regard to any enemy State if such State or any of its allies fails to respect the prohibitions laid down in the Protocol' (Source: SIPRI website available at <http://www.sipri.se/cbw/docs/cbw-hist-geneva-res.html#anchor479036>).
10. According to SIPRI, 'Biological anti-plant agents are definitely prohibited under the Geneva Protocol chemical herbicides were not discussed in 1925 (though biological anti-crop agents were referred to in the negotiations and were considered to be comprised under the Protocol's prohibition', see p.71

and 136, SIPRI, *The Problem of Chemical and Biological Warfare: CBW and the Law of War*, Vol. III, Humanities Press, New York, 1973.

11. J.G. Starke, QC offers the following definition of 'Protocol': 'This signifies an agreement less formal than a treaty or Convention proper and which is generally never in the Heads of State form. The term covers the following instruments: (a) An instrument subsidiary to a Convention, and drawn up by the same negotiators. Sometimes also called a Protocol of Signature, such a Protocol deals with ancillary matters such as the interpretation of particular clauses of the Convention, clauses not inserted in the Convention, or reservations by particular signatory States. Ratification of the Convention will normally *ipso facto* involve ratification of the protocol. (b) An ancillary instrument to a Convention, but of an independent character and operation and subject to independent ratification, for example, the Hague Protocols of 1930 on Statelessness, signed at the same time as the Hague Convention of 1930 on the Conflict of Nationality Laws. (c) An altogether independent treaty. (d) A record of certain understandings arrived at, more often called a *Proces-Verbal*. in *An Introduction to International Law*, 7th Edition, Butterworths London, 1972, p.403.
12. M. Dando, op. cit., p.35.
13. 'Special Conference of the States Parties to the Convention on the Prohibition on the Development, Production and Stockpiling of Bacteriological (Biological) and Toxin Weapons and on their Destruction, Final Report', BWC/SPCONF/1, Geneva, 19–30 September 1994.
14. Ibid.
15. Two additional Friends of the Chair were appointed at the July 1997 Ad Hoc Group meeting, one to address the Annex on Investigation (South Africa), and one to address Legal Issues (Australia). Two additional Friends of the Chair were appointed at the September 1997 Ad Hoc Group Meeting, one to address Confidentiality (Germany) and another to address Implementation and Assistance (India). See Tibor Toth, 'A Window of Opportunity for the BWC Ad Hoc Group', *The CBW Conventions Bulletin*, Issue No. 37, September 1997, p.2.
16. G.S. Pearson, 'The BTWC Enters the Endgame', *Disarmament Diplomacy*, Issue No. 39, July/ August 1999, p.13.
17. 'A List of Human, Animal and Plant Pathogens and Toxins', Second Ad Hoc Group Meeting, Annex III/8, Geneva, 10–21 July 1995.
18. 'Fourth Review Conference of the Parties to the Convention on the Prohibition of the Development, Production and Stockpiling of Bacteriological (Biological) and Toxin Weapons and on their Destruction, Final Document', P II, Final Declaration, 25 November to 6 December 1996, BWC/CONF.IV/9, Part II, p.15, paragraph 5.
19. G.S. Pearson, 'The BTWC Enters the Endgame', *Disarmament Diplomacy*, Issue No. 39, July/August 1999
20. T. Toth, 'Prospects for the Ad Hoc Group', in G.S. Pearson *et al.* (eds), *Verification of the Biological and Toxin Weapons Convention*, Kluwer Academic Publications, 1999, p.219.
21. *Plant Pathogens Important for the BWC*, Working Paper by South Africa, BWC/Ad Hoc Group/WP.124, 3 March 1997.

22. 'Ad Hoc Group of the States Parties to the Convention on the Prohibition of the Development, Production and Stockpiling of Bacteriological (Biological) and Toxin Weapons and on Their Destruction', Eighth Session, BWC/AD HOC GROUP/38, Rolling Text, Annex I, p.117.
23. 'Unlisted Pathogens Relevant to the Convention', Working Paper Submitted by Cuba, Seventh Session, BWC/AD HOC GROUP/WP.183, 22 July 1997.
24. 'Note to the Secretary General of the United Nations (UN)', circulated as UN General Assembly A/52/128, 29 April 1997.
25. Ambassador Ian Soutar, United Kingdom Permanent Representation to the Conference on Disarmament, 'Report of the Formal Consultative Meeting of States Parties to the Biological and Toxin Weapons Convention' held 25–27 August 1997, pp.1–3, G.E. 97-65258.

5 The Context of US BW Research and Development

1. Manfred Jonas, *Isolationism in America, 1935–1941*, Cornell University Press, Ithaca, New York, 1966, p.206.
2. Ibid.
3. David Dickson, *The New Politics of Science*, University of Chicago Press, 1988, p 25. See also, Carroll Pursell, *Science Agencies in WWII: The OSRD and Its Challengers*, in Nathan Reingold (ed.), *The Sciences in the American Context: New Perspectives*, Smithsonian Institution Press, Washington DC, 1979, pp.359–78.
4. Ibid., *p*.25.
5. In a discussion which treats scientific and technological innovation as what is referred to as a global so-called, 'independent variable', Barry Buzan, suggests that such innovation has a two-fold effect upon the process of armament: first, states are forced to assess their armament requirements not only in relation to the military capability of an adversary but also in relation to what Buzan refers to as the standard of 'technological leading edge', and second by, that the forward march of scientific and technological innovation conditioned the response of military planners which Buzan argues results in the inevitable and permanent institutionalisation of the domestic armament process. Barry Buzan, *An Introduction to Strategic Studies: Military Technology and International Relations*, Macmillan, IISS, 1987, p.127.
6. Carroll Pursell, 'Science Agencies in World War II: The OSRD and its Challengers', in Nathan Reingold (ed.), *The Sciences in the American Context: New Perspectives*, Smithsonian Institution Press, Washington, DC, 1979, p.359.
7. Other members of the Committee of the NDRC where, Conway p.Coe, as commissioner of patents, Richard D. Tolman, a Caltech professor, and one representative from the Army and one representative from the Navy. See, Carroll Pursell, ibid., *p*.360.
8. Ibid., p.361.
9. I. Stewart, *Organising Scientific Research for War: The Administrative History of the OSRD*, Little, Brown and Co, Boston, 1948, p.50.
10. David Dixon, op. cit., p.26.
11. Barton J Bernstein, 'The Birth of the US Biological-Warfare Program', *Scientific American*, Vol. 25, 6 June 1987, p.94.

12. Carroll Pursell, op. cit., p.364.
13. Ibid., p.375.
14. Ibid., *p*.375.
15. The prevailing War Department view of BW as impractical is thought to have been strongly influenced by Major Leon A. Fox, Chief of the Medical Section , US Chemical Warfare Service. See, *The Military Surgeon*, Vol. 72, No. 3, and Sheldon Harris, *Factories of Death*, p.150.
16. Ibid., *p*.152.
17. L.P. Brophy, W.D. Miles and R.C. Cochrane, *The Chemical Warfare Service: From Laboratory to Field*, Office of the Chief of Military History, Department of the Army, Washington, DC, 1959, p.103.
18. Ibid., *p*.103.
19. Letter from Secretary of War Stimson to Franklin D. Roosevelt,, 29 April 1942.
20. Report of the WBC Committee, 19 February 1942, p.3.
21. Barton J. Bernstein, 'Origins of the Biological Warfare Programme in the US, in Susan Wright', (ed.), *Preventing a Biological Arms Race*, The MIT Press, Cambridge, Mass., London, 1990, p.11.
22. Ibid., p.11.
23. SIPRI, *The Problem of Chemical and Biological Warfare: CB Weapons Today*, Vol. II, Humanities Press, New York, 1973, p.118.
24. Barton J. Bernstein, *Origins of the US Biological Warfare Program*, op. cit., p. 11.
25. Laughlin refers to the War 'Reserve' Service. See, L. L. Laughlin, U.S. Army Activity in the U.S. Biological Warfare Programs, Vol. I, 24 February 1977, p.I-I.
26. Economist Walter W. Stewart, Chairman of the Rockefeller Foundation, geographer Isaiah Bowman, president of Johns Hopkins University, and economist Edmund Ezra Day president of Cornell University, rejected offers to head the WRS. See, Barton J. Bernstein, *Origins of the US Biological Warfare Program*, op. cit., p.11.
27. Barton J. Bernstein, *The Birth of the U.S. Biological-Warfare Program*, op. cit., p.96.
28. Barton J. Bernstein, *Origins of the U.S. Biological Warfare Program*, op. cit., p. 13.
29. L.P. Brophy, W.D. Miles and R.C. Cochrane, op. cit., p.107.
30. According to Sheldon Harris, Vigo, 'if permitted to operate at full capacity, by 1945 was expected to be able to produce on a regular basis fully assembled 4-pound bombs filled with 4 percent anthrax slurry at 500,000 units per month. At the time of Japan's surrender, Vigo had on hand nearly 8000 pounds of the agent. The first shipments of bomb casing from the British company Electromaster were ready to be filled when orders were received to halt production. Those responsible for planning BW operation expected to use the anthrax bomb extensively. Orders to procure one million 4-pound bombs were placed with the British, with a delivery schedule of 125,000 per month beginning in March 1945. See, *Factories of Death: Japanese Biological Warfare 1932–1945 and the American Cover-up*, op cit., p.155.
31. L.P. Brophy, W.D. Miles and R.D. Cochrane, op. cit., p.36.
32. Ibid., p.102.

33. Ibid., *p*.111.
34. Ibid., *p*.102.
35. George W. Merck, Special Consultant for BW, '*Report to the Secretary of War*', 3 January 1946. Regarding the scope of US wartime investigations into BW, Merck went on to say that: 'All known pathogenic agents were subjected to thorough study and screening by scientists of the highest competence in their respective fields to determine the possibilities of such agents being used by the enemy', in L. L. Laughlin, *U.S. Army Activity in the U.S. Biological Warfare Programs,* Vol. II, 24 February 1977, p.A-3.
36. George W. Merck, ibid.
37. Ibid., *p*.D-2.
38. SIPRI, The Problem of Chemical and Biological Warfare: The Rise of CB Weapons, Vol. 1, Stockholm, 1971, p.122.
39. For a list of contracts set up by the War Research Service but terminated prior to 1 September 1944, and a list of contracts set up by the War Research Service and transferred to the Chemical Warfare Service on 1 July 1944 see: Leo L. Laughlin, op. cit., p.C2.
40. George W. Merck, op. cit., p.A-8.

6 Aspects of Anti-Crop BW Activity: France, Germany, Japan

1. S. Harris, *Factories of Death: Japanese Biological Warfare, 1932–45, and the American Cover-Up*, Routledge, New York, 1994, p.150.
2. B.J. Bernstein, 'The Birth of the U.S. Biological-Warfare Program', *Scientific American*, Vol. 256, June 1987, p.97. Although such preparations were indeed made, according to Franz *et al.*, the inoculation of 100,000 troops did not take place. See, J.L. Middlebrook and D.R. Franz, 'Botulinum Toxins', in D.R. Franz, E.T. Takafuji and F.R. Sidell, (eds), *Medical Aspects of Chemical and Biological Warfare*, Office of the Surgeon General, US Army, Falls Church, Virginia, 1997, p 644. This matter is discussed in some considerable detail by Donald Avery, 'Canadian Biological and Toxin Warfare Research, Development and Planning, 1925–45', in E. Geissler and J.E. van Courtland Moon (eds), *Biological and Toxin Weapons: Research, Development and Use from the Middle Ages to 1945*, SIPRI Chemical and Biological Warfare Study, No. 18, 1999, pp.190–1.
3. According to SIPRI, 'The UK maintained a substantial stockpile of chemical weapons until around 1957' *The Problem of Chemical and Biological Warfare: CB Weapons Today*, Vol. II, SIPRI, Humanities Press, New York, 1973, p.190.
4. 'A Review of German Activities in the Field of Biological Warfare', MIS, Intelligence Report, War Department, ALSOS Mission, Washington, DC, 12, September 1945, p 6.
5. M. Wheelis, 'Biological Sabotage in World War I', in E. Geissler and J.E. van Courtland Moon (eds), *Biological and Toxin Weapons: Research , Development and Use from the Middle Ages to 1945*, op. cit., p.54.
6. M. Hugh-Jones, Wicham Steed and German Biological Warfare Research, *Intelligence and National Security*, Vol. 7, No. 4 (1992), pp.379-402.
7. 'A Review of German Activities in the Field of Biological Warfare', op. cit., p. 114.

8. O. Lepick, 'French Activities Related to Biological Warfare: 1919–1945', in E. Geissler and J.E. van Courtland Moon (eds), *Biological and Toxin Weapons: Research, Development and Use from the Middle Ages to 1945*, op cit, pp. 70–80.
9. 'A Review of German Activities in the Field of Biological Warfare', op. cit., p. 8.
10. Ibid., p.114.
11. According to Goudsmit, the ALSOS Mission was, 'America's first serious effort in scientific intelligence'. See S.A Goudsmit, *Alsos, The History of Physics* 1800–1950, Tomash Publishers, 1988, p.X. Following the first Alsos Mission (Alsos I) that went into Italy in 1943 a subsequent, larger mission, sponsored by the US Army, the US Navy, General Grove's Intelligence Department and Vannevar Bush's Office of Scientific Research and Development followed the liberation of France. See Goudsmit, p.14.
12. 'A Review of German Activities in the Field of Biological Warfare', op. cit., p. 101.
13. See also E. Geissler, 'Biological Warfare Activities in Germany', in E. Geissler and J.E. van Courtland Moon (eds), *Biological and Toxin Weapons: Research, Development and Use from the Middle Ages to 1945*, SIPRI Chemical and Biological Warfare Studies, No. 18, p.117.
14. Ibid., p.104.
15. Ibid., p.105.
16. Ibid.
17. Ibid., p.10.
18. Ibid., p.132.
19. Germany unconditionally ratified the 1925 Geneva Protocol for the Prohibition of the Use in War of Asphyxiating, Poisonous or Other Gases, and of Bacteriological Methods of Warfare in 1929.
20. Sheldon Harris, 'Japanese Biological Warfare Research on Humans: A Case Study of Microbiology and Ethics', *Annals of the New York Academy of Sciences*, 1996, p.33.
21. According to Sheldon Harris, Ishii Shiro graduated with a medical degree from Kyoto Imperial University in 1920. He became Surgeon-First Lieutenant in the Japanese Imperial Army in 1921, and subsequently returned to Kyoto Imperial University where he obtained a PhD in microbiology in the late 1920s. He was promoted to the rank of major in 1930 and appointed Professor of Immunology at Tokyo Army Medical School where, according to Sheldon Harris, 'it became his self-assigned task to pursue biological warfare.' (see p.23). In late 1932 he established a laboratory at the city of Harbin in Japanese occupied Manchuria with a budget of 200,000 Yen. The then Lieutenant Colin Ishi Shiro was appointed Chief of the Kwantung Army's Water Purification Units in 1936. The designation of such units provided cover for biological warfare research and developments and of the 18 Water Purification Centres distributed throughout Japanese occupied Manchuria, Burma, Thailand and Singapore all are reported to have been involved in biological warfare research and development. Ibid. pp.23-26.
22. Sheldon Harris, *Factories of Death: Japanese Biological Warfare 1932–45 and the American Cover Up*, op. cit.

23. Report on Scientific Intelligence Survey in Japan, Volume V, 'Biological Warfare', September and October 1945, p.4.
24. Peter Williams and David Wallace, *Unit 731: the Japanese Army's Secret of Secrets*, St Edmundsbury Press Limited, 1989, p.75.
25. Ibid., p.76.
26. Ibid., p.69.
27. Ibid., p.101.

7 Post-Second World War US Anti-Crop BW

1. These figures vary in the literature. According to one source: 'One hundred and fifty six such papers were published between October 1945 and January 1947.' *The Problem of Chemical and Biological Warfare: Volume I, The Rise of CB Weapons*, SIPRI, 1971, p.121.
2. Leo L. Laughlin, *US Army Activity in the US Biological Warfare Programs*, Vol. 1, 24 February 1977, p.i.
3. CIA, *Quarterly Review of Biological Warfare Intelligence*, First Quarter, 1949, OSI/SR-1/49, p.3.
4. Note By The Secretaries to the Joint Intelligence Committee on 'Soviet Capabilities for Employing Biological and Chemical Weapons', J.I.G. 297/2, 27 January 1949, p.8.
5. Intelligence on Russian Development, Ministry of Defence, Defence Research Policy Committee, 18th Meeting, 6 December 1949, p.2.
6. Chiefs of Staff Committee, B.W. 'Report 1950–1951, Report by the Biological Warfare Sub-Committee', 12 March 1952, p.1.
7. H.I. Stubblefield, 'Resume of the Biological Warfare Effort', 21 March 1958, p.18.
8. SIPRI, op. cit., p.121.
9. H. I. Stubblefield, op. cit., p.28.
10. Miller outlines the chain of command from the respective services up to the Joint Chiefs of Staff as follows: 'Responsibility for war planning rested with the Joint Chiefs of Staff. Within the Joint Staff (organised under the Joint Chiefs of Staff) were two groups – the Strategic Plans Committee and the Strategic Plans Groups. (The latter was the working body.) Both groups were composed of representatives of the three services. Action to prepare strategic plans could be initiated by the Secretary of Defense, the Joint Chiefs of Staff, the Director of the Joint Staff, or the Joint Strategic Plans Committee. Normally, however, the initiating action came from the Joint Chiefs or the Joint Strategic Plans Committee. When war planning on the joint level had been prepared by the Strategic Plans Group, reviewed by the Strategic Plans Committee, and then approved by the Joint Chiefs of Staff, it became guidance for the Army, Navy, and Air Force for the preparation of strategic plans.' See D. L. Miller, *History of Air Force Participation in Biological Warfare Program, 1944–1951*, Historical Study 194, Wright Patterson Air Force Base, September 1952, part II, p.50.
11. 'Report on Special BW Operations', October 1948, p.7.
12. Ibid., p.7.
13. Ibid., p.8.

14. Special Operations Division and Central Intelligence Agency collaboration during the early Cold War years was the subject of investigation at Hearings before the Senate Select Committee on Governmental Operations with Respect to Intelligence Activities, September 1975. Leo L. Laughlin, op. cit., Vol. 1, pp.2–5.
15. Ibid., pp.2–3.
16. Chiefs of Staff Committee, op. cit., p.2.
17. 'Report of the Secretary of Defense's Ad Hoc Committee on Biological Warfare', 11 July 1949.
18. Ibid., p 1.
19. Ibid., p.4.
20. Ibid., p.6.
21. Ibid., p 8.
22. Ibid., p.9.
23. Ibid., p.10.
24. Joint Chiefs of Staff, 1837/8, 13 September 1949.
25. D.L. Miller, *Historical Study 194*, op. cit., p.38.
26. S. Wright, 'Evolution of Biological Warfare Policy, 1945–1990', in *Preventing a Biological Arms Race*, MIT Press, 1990.
27. The affirmation received National Security Council approval 16 February 1950 and was adopted by Truman 17 February 1950.
28. Earl P. Stevenson was the President of Arthur D. Little Company.
29. 'Report of the Secretary of Defense's Ad Hoc Committee on Chemical, Biological and Radiological Warfare', 30 June 1950, Appendix A, p.26.
30. Ibid., p.iii.
31. Ibid., p.iii.
32. Ibid., p.iv.
33. The Biological Department was re-designated the Biological Laboratories on 6 March 1951.
34. The Chemical Warfare Service was re-designated as the Chemical Corps 2 August 1946. See Leo p.Brophy, Wyndham D. Miles, and Rexmond C. Cochrane, *The Chemical Warfare Service: From Laboratory to Field*, Office of the Chief of Military History, Department of the Army, Washington, DC, 1959, p.22.
35. Colonel William M. Creasey, Chief, Research and Engineering Division, Office of the Chief, Chemical Corps, 'Report to the Secretary of Defense's Ad Hoc Committee on CEBAR', 24 February 1950, p.7, hereinafter referred to as the 'Creasey Report'.
36. 'Report of the Secretary of Defense's Ad Hoc Committee on Chemical, Biological and Radiological Warfare', op. cit., p.4.
37. Creasey Report, op. cit., p.7,
38. Ibid., p.7
39. Ibid.
40. Ibid., p.8
41. Ibid.
42. Ibid., p.9
43. Ibid.,
44. Ibid., p.11
45. Ibid.

46. 'Report of the Secretary of Defense's Ad Hoc Committee on Chemical, Biological and Radiological Warfare', op. cit., p.24.
47. Secretary of Defense, Memorandum for: The Secretary of the Army, The Secretary of the Navy, The Secretary of the Air Force, The Joint Chiefs of Staff, Chairman, Research and Development Board, Chairman, Munitions Board, Chairman, Military Liaison Committee, Director, Weapons Systems Evaluation Group, and Director of Public Information, Enclosure 'C', 27 October 1950. 'Report by the Joint Strategic Plans Committee to the Joint Chiefs of Staff on Chemical, Biological and Radiological Warfare', JCS 1837/18, pp.220-223.
48. Ibid.
49. Leo L. Laughlin, *U.S. Army Activity in the U.S. Biological Warfare Programs,* Vol. 1, op. cit., p. 3, ibid. pp. 3–4.
50. Ibid., pp. 3–4.
51. Research and Development Board Memoranda, RDB 64/44, 13 February 1948.
52. D.L. Miller, *History of Air Force Participation in Biological Warfare Programme, 1944–1951*, Historical Study N. 194, op. cit., p.33.
53. Leo p.Brophy, Wyndham D. Miles, and Rexmond C. Cochrane, op. cit., p. 110.
54. Joint Chiefs of Staff, 21 February 1951, Decision of J.C.S. 1837/18.
55. D.L. Miller, *History of Air Force Participation in Biological Warfare Programme, 1944–1951*, Historical Study No. 194, op. cit., p.5.
56. Ibid., p.13.
57. Ibid., p.55.
58. According to Miller the Advisory Weapons Systems Evaluation Group, which was composed of military and civilian specialists, was created 11 December 1948. Functioning under the Joint Chiefs of Staff and the Research and Development Board, the group analysed and evaluated weapons in the light of probable future combat conditions. D.L. Miller, *History of Air Force Participation in Biological Warfare Programme, 1944–1951*, Historical Study No. 194, op. cit.
59. Ibid., p.20.
60. Ibid., p.24.
61. Ibid., p.24.
62. Ibid., p.25.
63. Ibid., p.34.
64. Ibid., p.34.
65. Ibid., p.45.
66. D.L. Miller, *History of Air Force Participation in Biological Warfare Programme,* 1944–1951, Historical Study No. 303, Wright-Patterson Air Force Base, January 1957, Section 1, p.2.
67. Ibid., p.2.
68. D.L. Miller, *History of Air Force Participation in Biological Warfare Programme,* 1944–1951, Historical Study No. 194, op. cit., p.19.
69. D.L. Miller, History of Air Force Participation in Biological Warfare Programme, 1944- 1951, Historical Study Number 303, op. cit., p.69.
70. Ibid., p.70.
71. D.L. Miller, *History of Air Force Participation in Biological Warfare Programme, 1944–1951*, Historical Study No. 303, op. cit., p.6.

72. Ibid., p.53.
73. Ibid., p.61.
74. Leo L. Laughlin, op. cit., p.4-1.
75. Ibid., p.4-2.
76. Ibid., p.5-1.
77. SIPRI, *The Problem of Chemical and Biological Warfare: The Prevention of CBW*, Humanities Press, New York, Vol. V, 1971, pp.56–7.
78. Leo L. Laughlin, *U.S. Army Activity in the U.S. Biological Warfare Programs*, Vol. 2, 24 February 1977, p.D-3.
79. Ibid., p.D-3.

8 UK and US Anti-Crop Collaboration, 1943–58

1. G. Carter and G.S. Pearson, 'North Atlantic Chemical and Biological Research Collaboration: 1916–1995', *Journal of Strategic Studies*, Vol. 19, No. 1, March 1996, pp.74–103.
2. J. Bryden, *Deadly Allies*, McClelland & Stewart, Toronto, 1989.
3. SIPRI, *The Problem of Chemical and Biological Warfare*, Vols. I and II, Stockholm, Sweden, 1971 and 1973 respectively.
4. G. Carter and G.S. Pearson, op. cit., pp.74–103.
5. Where there has been a reliance on primary source documentation from official sources, such as, minutes of meetings and official reports, the narrative inevitably reflects, and is to some extent limited to, the construction of an institutional account of events.
6. For a detailed account of the origins of the British BW programme see, C.B. Carter and G.S. Pearson, 'British Biological Warfare and Biological Defence, 1925–45', in E. Geissler and J.E. van Courtland Moon (eds), *Biological and Toxin Weapons: Research, Development and Use from the Middle Ages to 1945*, SIPRI, Scorpion Series, No.18, p.168.
7. UK ratification of the Geneva Protocol was accompanied by reservations stating that it would be, 'bound only in respect of states parties and that the Protocol would cease to be binding in regard to states and their allies who failed to respect the prohibition'. See G.B.Carter and Graham S. Pearson, *British Biological Warfare and Biological Defence, 1925–45*, op. cit., p.168. See also A. Boserup, *The Problem of Chemical and Biological Warfare: CBW and the Law of War*, Vol. III, SIPRI, London, 1973, pp.78–9.
8. C.B. Carter and G.S. Pearson, op. cit., p.168.
9. Fildes research on aerosol took place prior to the emergence of commercial aerosols. According to Sanders, '[t]he first aerosol of any commercial significance, the aerosol insecticide, was defined in 1949 as a system of particles suspended in air where 80% of the particles were less than 30 microns in diameter and no particles were larger than 50 microns. This definition developed from observations that aerosol insecticides with this particle size were particularly effective because the particles remained suspended in air for approximately one hour'. See, p.Sanders, *Handbook of Aerosol Technology*, 2nd edn, London, 1979, p.3. One of the first patents for an aerosol was granted in 1862. The particle spray became smaller in the early 1900s, resulting in finer spray. During WWII,

the requirement for aerosolised insecticides for troops serving overseas was some 10,000 per day with one company (Westinghouse) supplying in excess of 30 million aerosols to the armed forces up to the end of the War (pp.6, 7).

10. Gradon Carter, 'Biological Warfare and Biological Defence in the United Kingdom, 1940–1979', *RUSI Journal*, December 1992, p.69.
11. Gradon Carter and Graham S. Pearson, 'North Atlantic Chemical and Biological Research Collaboration: 1916–1995', op. cit., p.82.
12. Gradon Carter, 'Biological Warfare and Biological Defence in the United Kingdom, 1940–1979', op. cit., p.70.
13. Gradon Carter and Graham S. Pearson, 'North Atlantic Chemical and Biological Research Collaboration: 1916–1995', op. cit., p.81.
14. Vannevar Bush, Office for Emergency Management, Office of Scientific Research and Development, letter to Sir John Anderson, 15 May 1944.
15. Ibid.
16. Letter from Sir John Anderson to Vannevar Bush , Director, Office for Emergency Management, Office of Scientific Research and Development (OSRD), 19 April 1944.
17. Sir John Anderson, Memorandum to the Prime Minister on Crop Destruction, 26 June 1944.
18. Memorandum to accompany letter from V. Bush under date of 15, May 1944, to Sir John Anderson.
19. Letter from Sir John Anderson to Vannevar Bush, op. cit.
20. Note by Joint Secretaries, 'Potentialities of Weapons of War During the Next Ten Years', Chiefs of Staff Committee, Joint Technical Warfare Committee, 12 November 1945. Enclosed Report by the Inter Services Sub-Committee on Biological Warfare.
21. Ibid.
22. Ibid.
23. Ibid.
24. Note by Joint Secretaries, 'Potentialities of Weapons of War During the Next Ten Years', op. cit.
25. Ibid.
26. Ibid.
27. Military Biology and Biological Warfare Agents, Department of the Army, TM 3-216, 1956.
28. B.W. Hanson, *Great Mistakes of the War*, New York, 1950, pp.88–107, re-printed in P.R. Baker, (ed.), *The Atomic Bomb: The Great Decision*, New York, 1968. Upon investigating this claim further Milton Leitenberg questioned the author directly to establish whether a correlation could be made between this claim and the following sentence which appeared on the first page (line 1 of paragraph 3) of the mimeographed press release, marked 'For Release at 7:30 P.M. E.S.T [Eastern Standard Time], 3 January 1946, Report to the Secretary of War by George W. Merck, Special Consultant for Biological Warfare': The sentence in Merck's press release read: 'Only the rapid ending of the war prevented field trials in an active theatre of synthetic agents which would, without injury to human or animal life, affect the growing crops and make them useless.' A correlation between Hanson's claim, and the Merck sentence referring to field trials of synthetic agents in an 'active theatre', if proved, would have sug-

gested that Truman might have secretly revoked Roosevelt's 'no-first-use' policy for chemical and biological weapons prior to the end of the war in the Pacific – a claim that was flatly denied by Truman (see Appendix 2). However, Hanson was unable to elaborate further regarding any such correlation (see correspondence to M. Leitenberg, Appendix II). Subsequent versions of the press release, and Merck's 'Report to the Secretary of War' have not included the sentence beginning: 'Only the rapid ending of the war'. See L.P.Brophy and G.J.B. Fisher, *The Chemical Warfare Service: Organizing for War, Office of the Chief of Military History*, United States Army, Washington, DC, 1959, p.107, note 42.

29. Reference to this Directive appeared some 5 years later in, Memorandum by the Chairman, B.W. Sub-Committee on BW Policy, Defence Research Policy Committee, Ministry of Defence, 11 May 1950, D.R.P. (50) 53, see note Ø, D.O. (45) 7th Meeting, Minute 5, at PRO, DEFE 10/ 26, 30087.
30. Ibid., p.1.
31. Brian Balmer, 'The Drift of Biological Weapons Policy in the UK 1945 – 1965', *Journal of Strategic Studies*, Vol. 20, No. 4, December 1997, p.117.
32. Biological Warfare Sub-Committee, 'Report on Biological Warfare', Annex to B.W. (47) 20 (Final), 8 September 1947, pp.27–8.
33. Ibid.
34. Gradon Carter and Brian Balmer, *Chemical and Biological Warfare and Defence*, 1945–1990, in R. Bud and P. Gummett (eds), *Cold War, Hot Science: Applied Research in Britain's Defence Laboratories*, 1945–1990, Harwood Academic Publishers, The Netherlands, 1999, pp.312–13.
35. Cabinet Defence Committee, 13th Meeting, (D(53), noted in Biological Warfare Report 1951 – September 1953, 8th December 1953, BW Sub-Committee, BW (53) 19, at PRO, WO188/666.
36. M.R. Dando, *Biological Warfare in the Twentieth Century*, Brassey's, London, 1994, p.49.
37. Brian Balmer, *The Drift of Biological Weapons Policy in the UK 1945–1965*, op. cit., p.118.
38. Ibid., p.118.
39. Ibid.
40. Ibid., p.125.
41. Ibid., p.128.
42. Ibid., p.129.
43. Gradon Carter and Brian Balmer, 'Chemical and Biological Warfare and Defence, 1945–1990', op. cit., p.315
44. Ibid., p.316.
45. According to the Ministry of Supply:, 'Broadly speaking, whilst there is considerable exchange of information on research and development there were few fields in which agreements exist on the allocation of effort between the two countries.' See, 'Note by the Chief Scientist', Ministry of Supply, Agreements Between US and UK on Allocation of Research and Development Effort, Defence Research Policy Committee, Ministry of Defence, 27 January, 1950, D.R.P. (50) 13, p.1, at PRO, DEFE 10/26, 23747.
46. Ibid., p.1.
47. Gradon Carter and Brian Balmer, 'Chemical and Biological Warfare and Defence, 1945–1990', op. cit., p.312.

48. Note by the Ministry of Supply, 'Review of the R. and D. Programmes – Chemical and Biological Warfare', Defence Research Policy Committee, Ministry of Defence, 8th October 1954, D.R.P. /P (54) 30, p.2, at PRO, DEFE 10/33.
49. Brian Balmer, *The Drift of Biological Weapons Policy in the UK 1945 – 1965*, op. cit., pp.115 – 45.
50. In 1950 the Secretary of the Defence Research Policy Committee noted that high level discussion should be held with the US Chiefs of Staff and Research and Development Board in order to discuss, 'the possibility of leaving this field of research and development [on BW] entirely to the Americans'. See Copy of Minute Dated 30 May 1950 from the Secretary, Defence Research Policy Committee to the Secretary Chiefs of Staff Committee, Note by the Joint Secretaries, 'Biological Warfare Policy,' Defence Research Policy Committee, Ministry of Defence, 9th June 1950, D.R.P. (50) 76 at PRO, DEFE 10/27, 30087.
51. First Meeting of the Crop Committee, Advisory Council on Scientific Research and Technical Development, Ministry of Supply, 10 August 1948, at PRO, WO195/9959.
52. Note by the Chief Scientist, Ministry of Supply, Advisory Council on Scientific Research and Technical Development, Committee on Crop Destruction, 21 January 1948, at PRO, WO195/9677.
53. Minutes, First Meeting of the Crop Committee, Ministry of Supply, 10 August 1948, op. cit.
54. 'Second Report of the Crop Committee', Ministry of Supply, 3 November 1948, at PRO, WO195, 10080.
55. 'A Note on Herbicide Research', Crop Committee, 1949, AC10031, at PRO, WO195, 10027.
56. 'A Note on the Water Content of Cereals', Crop Committee, 1948, AC10268, at PRO, WO195, 10264.
57. 'A Note on Spore Suspension Research, Crop Committee,' 1949, AC10372, at PRO, WO195, 10368.
58. 'A Note on the Use of Aircraft in Agriculture, Crop Committee,' 1949, AC10546, at PRO, WO195, 10542.
59. 'Investigations in to Anti-Crop Chemical Warfare', Crop Committee, 1949, AC10276, at PRO, WO195, 10272.
60. 'Countermeasures to Anti-Crop and Anti-Animal Warfare', Crop Committee, 1949, AC10742, at PRO, WO195, 10738.
61. 'Report on 2,4-D,' Crop Committee, 1950, AC10855, at PRO, WO195, 10851.
62. 'Investigation into Phytotoxicity', Crop Committee, 1950, AC11091, at PRO, WO195, 11087.
63. 'Effects of Insecticides and Herbicides on Animals', Crop Committee, 1951, AC11568, at PRO, WO195, 11564.
64. 'Anti Crop Detection and Destruction of Destructive Agents on Crop Leaf Samples', Crop Committee, 1952, 11728, at PRO, WO195, 11725.
65. 'Anti-Crop Aerial Spray Trials, Crop Committee', 1954, AC12929, at PRO, WO195, 12925.
66. 'Analysis of Rice Blast Epidemic Caused by *Piricularia oryzae',* Agricultural Defence Advisory Board, 1954, AC13199, at PRO, WO195, 13195.

67. 'Report on Cereal Rusts', Agricultural Defence Advisory Board, 1955, AC13378, at PRO, WO195, 13374.
68. 'Screening of *Piricularia* Isolates for Pathogenicity to Rice', Agricultural Defence Advisory Board, 1955, AC13381, at PRO, WO195, 13376.
69. 'Report on Clandestine Attacks on Crops and Livestock of the Commonwealth', Agricultural Defence Advisory Board, 1950, AC13154a, at PRO, WO195, 13150.
70. Extract from *Quarterly Report of the US Chemical Corps*, 1948, Crop Committee, AC10216, at PRO, WO195, 10212.
71. Extract from Quarterly Report of the US Chemical Corps, 1949, Crop Committee, AC10384, at PRO, WO195, 10380.
72. 'US Chemical Corps Report for Fiscal Year 1950', AC10761, at PRO, WO195, 10757.
73. 'Technical Summary Report', 4th Quarter, Camp Detrick, Crop Committee, 1950, AC 10862, at PRO, WO195, 10858.
74. Experiments at Camp Detrick on Plant Inhibitors on Sugar Beet Crops, Crop Committee, 1952, AC11737, at PRO, WO195, 11733.
75. Extract from US Chemical Corps Research and Development, 'Quarterly Progress Report', Crop Committee, 1953, AC12648, at PRO, WO195, 12644.
76. 'Secret Studies in Drying BW Agents', Extract from *Camp Detrick Technical Bulletin,* No.4, Crop Committee, AC12368, at PRO, WO195, 12364.
77. 'Report on Recent Work of the Crop Division at Camp Detrick', Agricultural Defence Advisory Board, 1955, AC13538, at PRO, WO195, 13534.
78. 'Report on the Processing, Packaging, and Storage of Cereal Rust Spores', Camp Detrick, Agricultural Defence Advisory Committee, 1955, at PRO, WO195, 13538.
79. 'Annual Review, 1955', Agricultural Defence Advisory Committee, Advisory Council on Scientific Research and Technical Development, Ministry of Supply, 14 November 1955 at PRO, WO195, 13473.
80. Ibid.
81. P.F. Cecil, *Herbicidal Warfare: The Ranch Hand Project in Vietnam*, Praeger, New York, 1986, p.17.
82. Allegations of use of BW agents by the US in Korea, however, were to persist for decades after the end of the war. See, Milton Leitenberg, Resolution of the Korean War Biological Warfare Allegations, *Critical Reviews in Microbiology*, 24, (3), 1998, pp.169–194.
83. P.F. Cecil, op. cit., p.18.
84. Ibid., p.19.
85. Minutes, Fourth Meeting of the Agricultural Defence Advisory Committee, Ministry of Supply, 1 March 1956, at PRO, WO195, AC13640.
86. G.D. Heath, 'A Note of the Recent Work of the Crops Division at Camp Detrick', Agricultural Defence Advisory Committee, Ministry of Supply, 21 December 1955 at PRO, WO195, A.C. 13538.
87. Brian Balmer, T*he Drift of Biological Weapons Policy in the UK 1945–1965*, op. cit., p.122. According to Balmer, 'Henderson had visited the US and expressed concern to BRAB in 1952 that American Colleagues of long standing had become offensively minded, wishing to end the Korean War as soon as possible.'

88. E.C. Tullis *et al.*, 'The Importance of Rice and the Possible Impact of Antirice Warfare', Technical Study 5, Office of the Deputy Commander for Scientific Activities, Biological Warfare Laboratories, Fort Detrick, Maryland, Fort Detrick Control Number 58-FDS-302.
89. G.D. Heath, 'A Note of the Recent Work of the Crops Division at Camp Detrick', *op. cit.*
90. Ibid., p.2.
91. For a list of Biological Field Testing between 1951 and 1969 of: Anti-Crop Pathogenic Agents Involving the Public Domain; (Unsubstantiated) Biological Field Testing of Anti-Crop Biological Agents Involving the Public Domain; and, Biological Field Testing of Anti-Crop Pathogenic Agents Not Involving the Public Domain see: Leo L. Laughlin, *US Army Activity in the US Biological Warfare Programs*, Vols 1 and 2 , 24 February 1977.
92. *'Secret Studies in Drying B.W. Agents'*, (in) Extracts from *Camp Detrick Technical Bulletin* No.4, op. cit., p.96.
93. A 'urediospore' is a spore of rust fungi.
94. Agrios offers the following explanation for 'obligate parasites': 'Of the large number of groups of living organisms, only a few members of a few groups can parasitize plants: fungi, bacteria, mycoplasmas, and parasitic higher plants, nematodes and protozoa, and viruses and viroids. These parasites are successful because they can invade a host plant, feed and proliferate in it, and withstand the conditions in which the host lives. Some parasites, include viruses, viroids, mycoplasmas, some fastidious bacteria, nematodes, and protozoa, and of the fungi those causing downy mildews, powdery mildews, and rusts are biotrophs, that is, they can grow and reproduce in nature only in living hosts, and they are called obligate parasites.' See G.N. Agrios, *Plant Pathology*, 3rd edn, Academic Press, Inc, London, 1998, p.41.
95. D.C.D.R.D. 'Cereal Rust – Processing, Packaging and Storage of Cereal Rust Spores', Agricultural Defence Advisory Committee, Advisory Council on Scientific Research and Technical Development, 30 December 1955, at PRO, WO195/13538.
96. Ibid.
97. 'Secret Studies in Drying B.W. Agents', Extracts from *Camp Detrick Technical Bulletin,* No.4, op. cit., p.104.
98. Ibid., p.105.
99. Specific strains of *Piricularia oryzae* and *Helminthosporium oryzae* used in this investigation were not identified by the authors.
100. 'Secret Studies in Drying B.W. Agents', (in) Extracts from *Camp Detrick Technical Bulletin,* No.4, op. cit., p.109.
101. Ibid., p.110.
102. Ibid., p.116.
103. Ibid., p.116.
104. Ibid., p.117.
105. Ibid., p.120.
106. Subsequent investigations into the effectiveness of *Phytophthora infestans* revealed in that its effectiveness could be enhanced. According to one report: 'A major breakthrough was achieved in 1956 by producing drought-resistant sexual oospores through mating of compatible strains of the fungus. Promising results have been obtained in initial attempts at

breaking the high degree of dormancy of the oospores, a characteristic which otherwise would limit seriously their operational employment.' See C.E. Minarik and Frances M. Latterell, 'Anti-Crop Agents', *BWL Technical Study 11*, Office of the Deputy Commander for Scientific Activities, Fort Detrick, Frederick, Maryland, 1958, Fort Detrick Control Number 58-FDS-666, p.vii.

107. 'Secret Studies in Drying B.W. Agents', in Extracts from *Camp Detrick Technical Bulletin,* No.4, op. cit., p.121.
108. Minutes, Fourth Meeting of the Agricultural Defence Advisory Committee, Ministry of Supply, 1 March 1956, at PRO, WO195, AC13640.
109. UK policy on chemical and biological warfare research and development in the early 1960s was governed by Chiefs of Staff guidance established by Chiefs of Staff document C.O.S. (58) 196, 1958. This policy is summarised as follows: 'a./ Neither BW nor CW had significant strategic value and there was negligible risk of the use of these methods against the UK, though outside the UK there was a risk of CW. There was no justification for defensive measures for UK based armed forces, for the civilian population or for ships at sea. b./ It was essential for the West as a whole to continue co-ordinated R and D. An offensive capability in BW and CW was essential, for the West, though not necessarily for the UK itself. The Armed forces still needed training against CW, and protective measures, in certain areas abroad. Limited training in BW and CW for the Civil Defence Services was visualised at the discretion of the Home Office. c./ The potential of BW and CW weapons was such that their military significance was to be reassessed periodically', in 'Chemical and Biological Warfare', Note by the Joint Secretaries, Defence Research Policy Committee (DRPC), Ministry of Defence, D.R.P./P (62) 33, 10 May 1962. At PRO, DEFE 10/490, 24144.

9 Munitions for Anti-Crop BW Agents

1. SIPRI, *The Problem of Chemical and Biological Warfare: CB Weapons Today*, Vol. II, 1973, pp.79–90.
2. J.P. Perry Robinson, 'Environmental Effects of Chemical and Biological Warfare', *War and Environment,* Environmental Advisory Council, Stockholm, Sweden, 1981, pp.1–43.
3. Assimilation: meaning that such weapons become integrated into the day-to-day routines, organisation, and prevailing doctrine of the armed forces. See J.P. Perry Robinson, Supply, Demand and Assimilation in Chemical Warfare Armament, in H. G. Brauch (ed.), *Military Technology, Armaments Dynamics and Disarmament,* Macmillan Press, 1989, p.112.
4. *Summary History of the Chemical Corps: 25 June 1950–8 September 1951*, Historical Office, Office Chief Chemical Officer, 30 October 1951, p.3.
5. D.L. Miller, *History of Air Force Participation in Biological Warfare Programme, 1944–1951,* Historical Study N 194, Wright-Patterson Air Force Base, September 1952, p.78. Hereinafter referred to as the 'Miller Report'.
6. Dorothy L. Miller, *History of Air Force Participation in Biological Warfare Program,* Historical Study No. 313, Wright-Patterson Air Force Base, January 1957, p.104, hereafter referred to as the *Miller Report.*

7. Testing with the E73 biological feather munition had revealed that in addition to anti-crop BW agents, the E73 feather munition could successfully disseminate the causal agent of hog cholera, or more significantly, according to Miller, 'a mixture' of anti-crop BW pathogens and anti-animal BW pathogens in the same munitions. See *Miller Report*, Historical Study, No. 313, op. cit., p.125.
8. *Miller Report*, Historical Study, No. 194, op. cit., p.79.
9. 'Feathers as Carriers of Biological Warfare Agents', Special Report, No. 138, Special Operation and Crops Division, 15 December, 1950. Declassified, 29 August 1977.
10. Ibid., p.2. The pathogen is described in the report as a, 'parasitic fungus of oats.'
11. Ibid.
12. Ibid., p.1.
13. Ibid., p.4.
14. Ibid., p.5.
15. Ibid., p.6.
16. *Miller Report*, Historical Study, No. 313, op. cit, p.105.
17. Ibid., p.106.
18. Ibid., p.110.
19. Leo L. Laughlin, *US Army Activity in the US Biological Warfare Programs*, Vol. 1, 24 February 1977, p.3-1.
20. R.C. Mikesh, *'Japan's World War II Balloon Bomb Attacks on North America'*, *Smithsonian Annals of Flight*, No. 9, Washington, 1973, and Bert Webber, Silent Siege III, Washington, 1992.
21. Ibid.
22. D.R. Franz, E.T.Takafuji, and F.R. Sidell, *Medical Aspects of Chemical and Biological Warfare*, Office of the Surgeon General, US Army, Falls Church, Virginia, p.51.
23. 'The Lighter than Air Force', *Air Force Magazine*, June 1999, p.65.
24. R.C. Mikesh, op. cit., p.9.
25. Ibid., p.22.
26. Ibid., p.22.
27. Ibid., p.17.
28. Ibid., p.9.
29. Ibid., p.21.
30. *Miller Report*, Historical Study No. 313, op. cit., p.112
31. Although hydrogen gas was not considered to be a safe as helium the latter of which was more expensive, the logistical problems associated with the extensive use of helium made its use impractical.
32. Kent Irish, *Anti-Crop Agent Munitions Systems*, BWL Technical Study 12, Annex B to Appendix VI, July 1958, of Operations Research Study Group, Study Number 21, U.S. Army Chemical Corps Operations Research Group (ORG) Army Chemical Center, Maryland, Volume 14, p.4, ORG-0511 (58), AD 347148, 1 August 1958.
33. Ibid., p.4.
34. Ibid., p.5.
35. Ibid., p.6.
36. *Miller Report*, Historical Study, No. 313, op. cit., p.116.

37. C.E. Minarik and Frances M. Latterell, *Anti-Crop Agents*, BWL Technical Study 11, August 1958, Office of the Deputy Commander for Scientific Activities, Fort Detrick, Frederick, Maryland, Fort Detrick Control Number 58-FDS-666, sub-section headed, 'The E77 foci coverage system was field-tested', p.vi.
38. *Miller Report*, Historical Study, No. 313, op. cit., p.125.
39. *Miller Report*, Historical Study, No. 313, op. cit., p.124. Anti-animal agents of greatest military interest were the causal agents of: fowl plague; foot and mouth disease; and, Rinderpest.
40. According to Miller, 'Balloons were being manufactured by General Mills Research, Inc., and the Wenzen Research Co., St. Paul, Minnesota.' The latter company was willing to expand for large-scale production. See Historical Study No. 313, op. cit., p.112.
41. According to Miller, 'The parsons contract involved two or three million dollars a year. Established in 1952, it was a continuing project. The company was handling five or six munition projects was also doing much work on research and new techniques and devices for forming aerosols. The company functioned at Camp Detrick under a project officer who was a civilian under the Civil Service.' See Historical Study No. 313, op. cit., p. 111.
42. Kent Irish, 'Anti-Crop Agent Munitions Systems', op. cit., p.4.
43. Ibid., pp.10–14. A design for a further dispenser for the dissemination of dry, undiluted biological anti-crop agents was proposed by General Mills. Some of the characteristics of this proposed unit, known as the Aero X3A, are described thus: 'The nose cone will contain the ram-air inlet and the tube to channel the air to the agent compartment. It also will cover the battery and motor-gear box assembly that is coupled to a paddle drive shaft. The centre section houses the agent container, which will have a payload capacity of 500 to 1500 pounds of agent. An agent agitation system, discharge orifice wiper and critical sharp-edged discharge orifice are parts of the centre section. The tail section maintains aerodynamic shape and supports the tail fins to stabilise the unit when jettisoned.'

10 Targets

1. Perkins, W.A., McMullen, R.W., Vaughan, L.M., 'Current Status of Anti-Crop Warfare Capability', Appendix VI, *Operational Effectiveness of Biological Warfare*, Operations Research Group Study N. 21, Vol. 13 (of 15 volumes), 1 August 1958, US Army Chemical Corps Operations Research Group, Army Chemical Center, Maryland.
2. E.C. Tullis *et al.*, *The Importance of Rice and the Possible Impact of Antirice Warfare*, Technical Study 5, Office of the Deputy Commander for Scientific Activities, Biological Warfare Laboratories, Fort Detrick, Maryland, Fort Detrick Control Number 58-FDS-302.
3. Perkins, W.A., McMullen, R.W., Vaughan, L.M., op. cit., p.14.
4. *The Importance of Rice and the Possible Impact of Antirice Warfare*, op. cit.
5. Ibid. Countries under this heading include: Burma, China, Formosa, India, Indo-China, Java, Japan, and Philippines.

6. Ibid., p.v.
7. A. Hay, 'A Magic Sword or a Big Itch: An Historical Look a the US Biological Warfare Programme', *Medicine, Conflict and Survival,* Vol., 15, 1999, pp. 215–234.
8. Ibid., pp.1–2.
9. Ibid., p.2.
10. D.H. Grist, Rice, Longmans, Green, London, New York, Toronto, 1953; T.H. Shen, *Agricultural Resources of China,* Cornell University Press, Ithaca, New York, 1951; J.G. DeGeus, *Means of Increasing Rice Production,* Centre D'Etude De L'Azote, Geneva, Switzerland, 1954; G.B. Cressey, *Land of the 500 Million,* McGraw-Hill, New York, Toronto, London, 1955; J.N. Efferson, 'The Story of Rice', *The Rice Journal,* Vol. 59, No. 7, Annual Issue, 1956; *Foreign Service Despatch,* No. 201, American Consulate General, Hong Kong, 1957.
11. *The Importance of Rice and the Possible Impact of Antirice Warfare,* op. cit., p. 5.
12. Ibid., p.6.
13. Ibid., p.7.
14. Ibid., p.9.
15. Ibid.
16. Ibid.
17. Ibid., p.10.
18. Ibid.
19. Ibid., p.11.
20. Ibid., p.12.
21. Ibid., p.14.
22. Ibid., p.14.
23. Ibid., p.16.
24. Ibid., p.18.
25. Ibid., p.19.
26. Ibid., p.20.
27. Ibid., p.20.
28. Ibid., p.21.
29. Ibid., p.46.
30. Ibid., p.50.
31. Ibid., p.67.
32. Ibid., p.74.
33. Ibid., p.80.
34. Ibid., p.81.
35. Ibid., p.91.
36. Ibid., p.82.
37. Ibid., p.88.
38. Ibid., p.88.
39. Ibid., p.94.
40. Ibid., pp.97–8.
41. Ibid., p.100.
42. Ibid., p.102.
43. Ibid., p.102
44. Ibid., p.103.
45. Ibid., p.103.

46. Ibid., p.104.
47. Ibid., p.105.
48. Ibid., p.106.
49. Ibid., p.111.
50. Ibid., p.112.
51. Ibid., p.114.
52. Ibid., p.115.
53. Ibid., p.117. See Gaumann, *Principles of Plant Infection*, Hafner, New York, 1951.
54. *The Importance of Rice and the Possible Impact of Antirice Warfare*, op. cit., p.117.
55. Ibid., p.119.
56. Ibid., p.119.
57. Ibid., p.119.
58. Ibid., p.118.
59. Ibid.
60. Ibid., p.121.
61. Ibid., p.129.
62. Ibid., p.129.
63. Ibid., p.144. The Census for Communist China 1953 (1 November 1954), estimated that the population for mainland China was 582,602,477 (corrected to 582,603,417. This figure has been rounded to 500 million in the report.
64. Ibid., p.129.
65. Ibid., p.130.
66 Ibid., p.149.
67. Ibid., p.154.
68. Ibid., p.123.
69. Ibid., p.150.
70. Ibid., p.159.
71. Ibid., p.161.
72. Ibid., p.154. Double-cropping refers to growing two successive crops in the same season on the same land, the second crop put on the land after the first had been removed.
73. Ibid., p.178.
74. Ibid., p.000.
75. *Biological Agents and Munitions Data Book*, Miscellaneous Publication 34, Systems Analysis Division, Department of the Army, Fort Detrick, Maryland, November, 1969.

11 Conclusions

1. U. Deichmann, *Biologists Under Hitler*, Harvard University Press, Massachusetts, 1996, p.280.
2. 'Combating Proliferation of Weapons of Mass Destruction', Report of the Commission to Assess the Organization of the Federal Government to Combat the Proliferation of Weapons of Mass Destruction, July 1999, p.91.
3. Ibid.
4. Ibid.

5. M.R. Dando, 'Technological Change and Future Biological Warfare', paper presented at a conference on 'Biological Warfare and Disarmament: Problems, Perspectives, Possible Solutions', UNIDIR, Geneva, 5–8 July, 1995, p.5.
6. M. Wheelis, 'Biological Sabotage Operations in World War I', in E. Geissler and J.E. van Courtland Moon (eds), *Biological and Toxin Weapons: Research, Development and Use from the Middle Ages to 1945*, SIPRI Chemical and Biological Warfare Studies No. 18, Oxford University Press, 1999.
7. O. Lepick, 'French Activities Related to Biological Warfare, 1919–45', in E. Geissler and J.E. van Courtland Moon (eds), *Biological and Toxin Weapons: Research, Development and Use from the Middle Ages to 1945*, op. cit.
8. J.A. Poupard and L.A. Miller, 'History of Biological Warfare: Catapults to Capsomeres', *Annals of the New York Academy of Sciences*, Vol. 666, 1992, p.14.
9. See: C.S. Cox, *The Aerobiological Pathway of Microorganisms*, John Wiley, Chichester, 1987.
10. *The Problem of Chemical and Biological Warfare: CB Weapons Today*, SIPRI, Humanities Press, New York, 1973, p.128.
11. E. Geissler, 'A New Generation of Biological Weapons', in E. Geissler (ed.), *Biological and Toxin Weapons Today*, SIPRI, Oxford University Press, 1986, p.22.
12. *The Problem of Chemical and Biological Warfare: CB Weapons Today*, op. cit., pp.82–9.
13. Malcolm Dando, 'Biological Weapons: Curbing the Test Tube Danger', *New Zealand International Review*, January/February, 1994.
14. Ibid.
15. E. Geissler, *A New Generation of Biological Weapons*, op. cit., p.22.
16. Ibid., p.23.
17. K. Alibek, *Biohazard*, Hutchinson, London, 1999, p.281.
18. T.G. Roetzel, *Biological Agents and Munitions Data Book*, Miscellaneous Publication 34, Department of the Army, Fort Detrick, Frederick, Maryland, Fort Detrick Control Number, 69-FDS-1041, 1969 p.82 (see Appendix III).
19. W. Cohen, *Proliferation: Threat and Response*, U.S. Department of Defense, Washington, DC, 1997, Technical Annex, 'Responding to Novel Biological Warfare Threats', available at: <http://www.defenselink.mil/pubs/prolif97/annex.html#technical>
20. A literature search of agriculturally related life science databases (including: the BIDS, Science Citation Index; EDINA, the Commonwealth Agricultural Bureau Database, and GEOBASE, Electronic Reference Library Database) conducted in November 1999 on the terms 'genome of rice' and '*Piricularia oryzae*'(the causal agent of rice blast); and on 'genome of wheat' and '*Puccinia graminis*' (the causal agent of wheat rust) resulted in 108 agriculturally related genome study citations (as defined by the above terms).
21. M. Meselson, *The Problem of Biological Weapons*. Presentation given at the 1818th Stated Meeting of the American Academy of Arts and Sciences, Cambridge, 13 January 1999. Subsequently presented at Pugwash Meeting No. 246, 11th Workshop of the Pugwash Study Group on the Implementation of the Chemical and Biological Weapons Conventions: Implications of the CWC Implementation for the BTWC Protocol Negotiations Norrdwijk, The Netherlands, 15–16 May 1999.

22. G.S. Pearson. 'The Vital Importance of the Web of Deterrence', Proceedings, Sixth International Symposium on Protection Against Chemical and Biological Warfare Agents, Stockholm, 11–15 May 1998.

12 Related International Regulations

1. 'Convention on the Prohibition of the Development, Production and Stockpiling of Bacteriological (Biological) and Toxin Weapons and on Their Destruction', 1972, Article X.
2. 'Final Declaration, Fourth Review Conference of the Parties to the Convention on the Prohibition of the Development, Production and Stockpiling of Bacteriological (Biological) and Toxin Weapons and on their Destruction, Final Document', Part II, BWC/CONF. IV/9, 1996, Part II, page 23.
3. P. Rogers, S. Whitby, and M. Dando, 'Germ War Against Crops', *Scientific American*, Vol. 280, No. 6, June 1999, p.67.
4. Tansey, G, (1999) Trade, 'Intellectual Property, Food and Biodiversity', *A Discussion Paper*, Quaker Peace and Service, pp.1–24.
5. 'Convention on Biological Diversity', opened for signature at Rio de Janeiro 5 June 1992, HMSO, Cm 2127, January 1993. Also available at United Nations, UNEP/CBD/94/1, Geneva, November, Article I.
6. Ibid.
7. United Nations (1995) Convention on Biological Diversity, Second Conference, Jakarta, 1995.
8. United Nations. United Nations Environment Programme (UNEP), *UNEP International Technical Guidelines for Safety in Biotechnology,* UNEP Nairobi, Kenya, I (Introduction), 1995.
9. International Plant Protection Convention. The 1979 Convention is available at <http://www.fao.org/ag/agp/agpp/pq/Conven/1979text.htm>.
10. International Plant Protection Convention. The New Revised Text approved by the FAO Conference at its 29th Session, November 1997 is available at <www.fao.org/ag/agp/agpp/pq/Conven/conventn.htm>.
11. Wheelis, M., Outbreaks of Disease: Current Disease Reporting, *Briefing Paper* No.21, April 1999, p.1–16.
12. Pearson, G.S., Article X: Further Building Blocks, *Strengthening the Biological Weapons Convention, Briefing Paper No.7,* March, 1998, p.32.
13. G.S. Pearson and N. Sims, National Implementation Measures, *Strengthening the Biological and Toxin Weapons Convention, Briefing Paper No.4,* January 1998.
14. ProMED-mail is available at <http://www.healthnet.org/programs/promed.html> Pearson, G.S., Article X: Some Building Blocks, 'Strengthening the Biological Weapons Convention', Briefing Paper No.6, March, 1998, p.1.
15. Pearson, G.S., Article X: Some Building Blocks, 'Strengthening the Biological Weapons Convention', Briefing Paper No.6, March, 1998, p.1.
16. Rogers, P., Whitby, S., and Dando, M., 'Biological Warfare Against Crops', *Scientific American*, Vol. 20, No.6, June 1999.

Appendix 1

1. For a more detailed discussion on the blurring of the distinction between offensive and defensive BW research and development, see: Harlee Strauss

and Jonathan King, 'The fallacy of defensive biological weapon programmes', in E. Geissler (ed.), *Biological and Toxin Weapons Today*, SIPRI, Oxford University Press, 1986.
2. George N. Agrios, *Plant Pathology*, 3rd edn, Academic Press, 1988, p.359.
3. According to Agrios, 'mycelium' refers to the body of the fungus, Ibid., p.265.
4. Arden F. Sherf, Robert M. Page, E.C. Tullis, and Thomas L. Morgan, 'Studies on Factors Affecting the Infectivity of *Helminthosporium Oryzae*', *Phytopathology*, Vol. 37, No. 281, May 1947, p.282.
5. Ibid., p.282.
6. Ibid., p.286.
7. Ibid., p.288.

Bibliography

1 Books

Agrios, George N. *Plant Pathology*, 3rd ed., Academic Press, New York, 1988.

Alibek. K. *Biohazard*, Hutchinson, London, 1999.

Baker, P.R. (ed.). *The Atomic Bomb: The Great Decision*, New York, 1968.

Beaumont, P., Blake, G.H. and Wagstaff, J.M. (eds). *The Middle East: A Geographical Study*, 2nd edn, David Fulton, London, 1988.

Brophy, L.P. and Fisher, G.J.B. *The Chemical Warfare Service: Organizing for War*, Office of the Chief of Military History, United States Army, Washington DC, 1959.

Brophy, L.P., Miles, W.D. and Cochrane, R.C. *The Chemical Warfare Service: From Laboratory to Field*, Office of the Chief of Military History, Department of the Army, Washington, DC, 1959.

Bryden, J. *Deadly Allies*, McClelland & Stewart, Toronto, 1989.

Buzan, B. *An Introduction to Strategic Studies: Military Technology and International Relations*, Macmillan, IISS, 1987.

Cecil, P.F. *Herbicidal Warfare: The Ranch Hand Project in Vietnam*, Praeger, New York, 1986.

Chapman, S.R. and Carter, L.P. *Crop Production: Principles and Practices*, W.H. Freeman, San Francisco, 1975.

van Courtland Moon, J.E. 'Controlling Chemical and Biological Weapons Through World War II', in R.D Burns (ed.), *Encyclopaedia of Arms Control and Disarmament*, Vol. II, Charles Schribner's Sons, New York, 1993.

Covert, N. *Cutting Edge: A History of Fort Detrick, Maryland – 1943–1993'* Headquarters, United States Army Garrison, Fort Detrick, Maryland.

Cox, C.S. *The Aerobiological Pathway of Microorganisms*, John Wiley, Chichester, 1987.

Cressey, G.B. *Land of the 500 Million*, McGraw-Hill, New York, Toronto, London, 1955.

Dando, M.R. *Biological Warfare in the Twentieth Century*, Brassey's, London, 1994.

Dando, M.R. *Biotechnology, Weapons and Humanity*, Harwood Academic Publishers, (for the British Medical Association), 1999.

DeGeus, J.G. *Means of Increasing Rice Production*, Centre D'Etude De L'Azote, Geneva, Switzerland, 1954.

Deichmann, U. *Biologists Under Hitler*, Harvard University Press, Massachusetts, 1996.

Dickson, D. *The New Politics of Science*, University of Chicago Press, 1988.

Franz, D.R. Takafuji, E.T. and Sidell, F.R. Medical Aspects of Chemical and Biological Warfare, Office of the Surgeon General, US Army, Falls Church, Virginia, 1994.

Gaumann, E. *Principles of Plant Infection*, Hafner, New York, 1951.

Geissler, E. 'A New Generation of Biological Weapons, in E. Geissler (ed.), *Biological and Toxin Weapons Today*, SIPRI, Oxford University Press, 1986.

Grist, D.H. *Rice*, Longmans, Green, London, New York, Toronto, 1953.
Goudsmit, S.A. *Alsos, The History of Physics* 1800–1950, Tomash Publishers, 1988.
Hanson, B. W. *Great Mistakes of the War*, New York, 1950.
Jonas, M. *Isolationism in America, 1935–1941*, Cornell University Press, Ithaca, New York, 1966.
Harris, S. Factories of Death: Japanese Biological Warfare, 1932–45, and the American Cover-Up, Routledge, New York, 1994.
Hugo and Russell (eds), *Pharmaceutical Microbiology*, 6th edn, 1998.
Mikesh, R.C. 'Japan's World War II Balloon Bomb Attacks on North America', Smithsonian Annals of Flight, No. 9, Washington, 1973.
Oerke, E.C. Dehne, H.W. Schonbeck F. and Weber, A. *Crop Production and Crop Protection: Estimated Losses in Major Food and Cash Crops*, Elsevier, Oxford, 1994.
Parry, D.W. *Plant Pathology in Agriculture*, Cambridge University Press, 1989.
Pearson, G.S. *The UNSCOM Saga: Lessons for Countering the Proliferation of Chemical and Biological Weapons*, Macmillan 1999.
van der Plank, J.E. *Plant Diseases: Epidemics and Control*, Academic Press, New York, 1963
Rosebury, T. *Peace or Pestilence*, McGraw-HIll, New York, 1949.
Sanders, P. Handbook of Aerosol Technology, 2nd edn, London, 1979.
Shen, T.H. *Agricultural Resources of China*, Cornell University Press, Ithaca, New York, 1951.
SIPRI Yearbook, World Armaments and Disarmament, Oxford University Press, 1989.
Starke, J.G. Q.C. *An Introduction to International Law*, 7th edn, Butterworth, London, 1972.
Stewart, I. *Organising Scientific Research for War: The Administrative History of the OSRD*, Little, Brown, Boston, 1948.
The Problem of Chemical and Biological Warfare: The Rise of CB Weapons, Vol. 1, SIPRI, Humanities Press, New York, 1971.
The Problem of Chemical and Biological Warfare: CB Weapons Today, Vol. 1, SIPRI, Humanities Press, New York, 1973.
The Problem of Chemical and Biological Warfare: CBW and the Law of War, Vol. III, SIPRI, Humanities Press, New York, 1973.
The Problem of Chemical and Biological Warfare: CB Disarmament Negotiations: 1920–1970, Vol. IV, SIPRI, Humanities Press, New York, 1971.
The Problem of Chemical and Biological Warfare: The Prevention of CBW, Vol. V, SIPRI, Humanities Press, New York, 1971.
The Problem of Chemical and Biological Warfare: Technical Aspects of Early Warning and Verification, Vol. VI, 1973.
Wright, S. (ed.) *Preventing a Biological Arms Race*, MIT Press, 1990.
Stakman, E.C. and Harrar, J.G. *Principles of Plant Pathology*, The Ronald Press, New York, 1957.
Webber, B. *Silent Siege III*, Washington, 1992.
Williams, P. and Wallace, D. *Unit 731: the Japanese Army's Secret of Secrets*, St Edmundsbury Press, 1989.

2 Chapters in books

Avery, D. 'Canadian Biological and Toxin Warfare Research, Development and Planning, 1925–45, in Geissler, E. and van Courtland Moon, J.E. (eds),

Biological and Toxin Weapons: Research, Development and Use from the Middle Ages to 1945, SIPRI Chemical and Biological Warfare Study, No. 18, 1999

Bernstein, B.J. 'Origins of the Biological Warfare Programme in the US', in Susan Wright, (ed.), *Preventing a Biological Arms Race*, The MIT Press, Cambridge, Mass., London, 1990.

Carter, G. and Balmer, B. 'Chemical and Biological Warfare and Defence, 1945–1990', in R. Bud and P. Gummett (eds), *Cold War, Hot Science: Applied Research in Britain's Defence Laboratories*, 1945–1990, Harwood Academic Publishers, The Netherlands, 1999.

Carter, C.B. and Pearson, G.S. 'British Biological Warfare and Biological Defence, 1925–45', in Geissler, E. and van Courtland Moon, J.E. (eds), *Biological and Toxin Weapons: Research, Development and Use from the Middle Ages to 1945*, SIPRI, Scorpion Series, No.18.

Dando, M.R. 'The Development of International Legal Constraints on Biological Warfare in the 20th Century', *The Finnish Yearbook of International Law*, Kluwer International, 1999.

Geissler, E. van Courtland Moon, J.E. and Pearson, G.S. 'Lessons from the History of Biological and Toxin Warfare', in Geissler, E. and van Courtland Mood, J.E. (eds), *Biological and Toxin Weapons: Research, Development and Use from the Middle Ages to 1945*, SIPRI Chemical and Biological Warfare Studies No. 18, Oxford University Press, 1999.

Lepick, O. 'French Activities Related to Biological Warfare: 1919–1945, in Geissler, E. and van Courtland Moon, J.E. (eds), *Biological and Toxin Weapons Research, Development and Use from the Middle Ages to 1945: A Critical Comparative Analysis*, Oxford University Press, SIPRI Study No.18, 1999.

Middlebrook, J. L. and Franz, D. R. 'Botulinum Toxins', in Franz, D.R. Takafuji, E.T. and Sidell, F.R. (eds), *Medical Aspects of Chemical and Biological Warfare*, Office of the Surgeon General, US Army, Falls Church, Virginia, 1997.

Perry Robinson, J.P. *Environmental Effects of Chemical and Biological Warfare*, War and Environment, Environmental Advisory Council, Stockholm, Sweden, 1981.

Perry Robinson, J.P. 'Supply, Demand and Assimilation in Chemical Warfare Armament', in H.G. Brauch (ed.) *Military Technology, Armaments Dynamics and Disarmament*, Macmillan Press, 1989.

Pursell, C. 'Science Agencies in WWII: The OSRD and its Challengers', in Reingold, N. (ed.), *The Sciences in the American Context: New Perspectives*, Smithsonian Institution Press, Washington, DC, 1979.

Toth, T. 'Prospects for the Ad Hoc Group', in G.S. Pearson et al. (eds), *Verification of the Biological and Toxin Weapons Convention*, Kluwer Academic Publications, 1999.

Wheelis, M. 'Biological Warfare Before 1914: The Prescientific Era', in Geissler, E. and van Courtland Moon, J.E. (eds), *Biological and Toxin Weapons Research, Development and Use from the Middle Ages to 1945: A Critical Comparative Analysis*, SIPRI Study No.18, Oxford University Press, 1999.

Wheelis, M. 'Biological Sabotage in the First World War', in E. Geissler and J.E. van Courtland Moon (eds), *Biological and Toxin Weapons Research, Development and Use from the Middle Ages to 1945: A Critical Comparative Analysis*, Oxford University Press, SIPRI Study No.18, 1999.

Wright, S. 'Evolution of Biological Warfare Policy, 1945–1990', in *Preventing a Biological Arms Race*, MIT Press, 1990

3 Journal articles

Andersen, A.L. Henry, B.W. and Tullis, E.C. 'Factors Affecting Infectivity, Spread, and Persistence of *Piricularia Oryzae* Cav', *Phytopathology*, Vol. 37, February 1947.

Balmer, B. 'The Drift of Biological Weapons Policy in the UK 1945–1965', *Journal of Strategic Studies*, Vol. 20, No. 4, December 1997.

Bernstein, B.J. 'The Birth of the US Biological-Warfare Program', *Scientific American*, Vol. 25, 6 June 1987.

Carter, G. 'Biological Warfare and Biological Defence in the United Kingdom, 1940–1979', *RUSI Journal*, December 1992.

Carter, G. and Pearson, G.S. 'North Atlantic Chemical and Biological Research Collaboration: 1916–1995', *Journal of Strategic Studies*, Vol. 19, No. 1, March 1996.

Dando, M. 'Biological Weapons: Curbing the Test Tube Danger', *New Zealand International Review*, January/February, 1994.

Efferson, J.N. 'The Story of Rice', *The Rice Journal*, Vol. 59, No. 7, Annual Issue, 1956.

Harris, S. 'Japanese Biological Warfare Research on Humans: A Case Study of Microbiology and Ethics', *Annals of the New York Academy of Sciences*, 1996.

Hay, A. 'A Magic Sword or a Big Itch: An Historical Look at the US Biological Warfare Programme', *Medicine, Conflict and Survival*, Vol. 15, 1999.

Kadlec, Lt Col. R.P. 'Biological Weapons for Waging Economic Warfare, in Battlefield of the Future: 21st Century Warfare Issues', <http://www.cdsar.af.mil/battle/bftoc.html>

Leitenberg, M. 'Resolution of the Korean War Biological Warfare Allegations', *Critical Reviews in Microbiology*, 24, (3), 1998.

Page, R.M. Tullis, E.C. and Morgan, T.L. 'Studies on Factors Affecting the Infectivity of Helminthosporium Oryzae', *Phytopathology*, 37: 281, May, 1947.

Page, R.M., Sherf A.F., and Morgan, T.L. 'The Effect of Temperature and Relative Humidity on the Longevity of the Conidia of *Helminthosporium oryzae*', *Mycologia*, Vol. 49, March-pril, 1947.

Pearson, G.S. 'The BTWC Enters the Endgame', *Disarmament Diplomacy*, Issue No. 39, July/ August 1999.

Pearson, G.S. 'The BTWC Enters the Endgame', *Disarmament Diplomacy*, Issue No. 39, July/August 1999.

Perry Robinson, J.P. 'Some Political Aspects of the Control of Biological Weapons', *Science in Parliament*, Vol. 53, No. 3, May/June, 1996.

Poupard J.A., and Miller, L.A. 'History of Biological Warfare: Catapults to Capsomers', *Annals of the New York Academy of Sciences*, 1992.

Rogers, P., Whitby, S., and Dando, M. Germ War Against Crops, *Scientific American*, Vol. 280, No. 6, June 1999.

Sherf, Arden F. Page, R.M. Tullis, E.C. and Morgan, T.L. 'Studies on Factors Affecting the Infectivity of *Helminthosporium Oryzae, Phytopathology*', Vol. 37, No. 281, May 1947.

Snieszko, S.F. Carpenter, J.B. Lowe, E.P. and Jakob, J.G. 'Improved Methods for the Cultivation and Storage of *Phytophthora infestans*', *Phytopathology*, Vol. 37, No. 635, September 1947.

Toth, T. 'A Window of Opportunity for the BWC Ad Hoc Group', *The CBW Conventions Bulletin*, Issue No. 37, September 1997.

Tucker, J. 'Strengthening the BWC: Moving Toward a Compliance Protocol', *Arms Control Today*, January/February, 1998.
Zilinskas, R.A., 'Iraq's Biological Weapons: The Past as Future', *Journal of the American Medical Association*, Vol. 278, No.5, August 6, 1997.

4 Official documentation and United Nations reports

'A List of Human, Animal and Plant Pathogens and Toxins', Second Ad Hoc Group Meeting, Annex III/8, Geneva, 10–21 July 1995.
Ad Hoc Group of the States Parties to the Convention on the Prohibition of the Development, Production and Stockpiling of Bacteriological (Biological) and Toxin Weapons and on Their Destruction, Eighth Session, 1972 BWC/AD HOC GROUP/38.
Ambassador Ian Soutar, United Kingdom Permanent Representation to the Conference on Disarmament, Report of the Formal Consultative Meeting of States Parties to the Biological and Toxin Weapons Convention held 25–27 August 1997, pp. 1–3, G.E. 97-65258.
W. Cohen, *Proliferation: Threat and Response*, US Department of Defense, Washington, DC, 1997, p. Technical Annex, Responding to Novel Biological Warfare Threats, available at: http://www.defenselink.mil/pubs/prolif97/annex.html#technical>.
'Combating Proliferation of Weapons of Mass Destruction', Report of the Commission to Assess the Organization of the Federal Government to Combat the Proliferation of Weapons of Mass Destruction, July 1999.
'Convention on Biological Diversity', opened for signature at Rio de Janeiro, 5 June 1992, HMSO, Cm 2127, January 1993. Available at United Nations, UNEP/CBD/94/1, Geneva.
'Convention on the Prohibition of the Development, Production and Stockpiling of Bacteriological (Biological) and Toxin Weapons and on Their Destruction', United Nations, 1972.
'Final Declaration, Fourth Review Conference of the Parties to the Convention on the Prohibition of the Development, Production and Stockpiling of Bacteriological (Biological) and Toxin Weapons and on their Destruction, Final Document', Part II, BWC/CONF. IV/9, 1996.
'Fourth Review Conference of the Parties to the Convention on the Prohibition of the Development, Production and Stockpiling of Bacteriological (Biological) and Toxin Weapons and on their Destruction. Final Document', Part II, Final Declaration, 25 November to 6 December 1996, BWC/CONF.IV/9, Part II.
'Geneva Protocol for the Prohibition of the Use in War of Asphyxiating, Poisonous or other Gases, and of Bacteriological Methods of Warfare', Signed at Geneva on 19 June 1925.
'International Plant Protection Convention', the 1979 Convention is available at <http://www.fao.org/ag/agp/agpp/pq/Conven/1979text.htm>.
'International Plant Protection Convention', the New Revised Text approved by the FAO Conference at its 29th Session, November 1997 is available at <www.fao.org/ag/agp/agpp/pq/Conven/conventn.htm>.
Note to the Secretary General of the United Nations, circulated as UN General Assembly A/52/128, 29 April 1997.

'Plan for Future Ongoing Monitoring and Verification of Iraq's Compliance with Relevant Parts of Section C of Security Council Resolution 687 (1991)', S/22871/Rev.1, 2 October 1991.
'Plant Pathogens Important for the BWC', Working Paper by South Africa, BWC/Ad Hoc Group/WP.124, 3 March 1997.
'Special Conference of the States Parties to the Convention on the Prohibition on the Development, Production and Stockpiling of Bacteriological (Biological) and Toxin Weapons and on their Destruction, Final Report', BWC/SPCONF/1, Geneva, 19–30 September 1994.
United Nations. *United Nations UNEP International Technical Guidelines for Safety in Biotechnology, United Nations Environment Programme (UNEP),* UNEP Nairobi, Kenya, I (Introduction), 1995.
United Nations Security Council Resolution 687, S/RES/687/8 April 1991, adopted by the Security Council at its 2,981st meeting, on 3 April 1991.
United Nations Security Council Resolutions: 660, 661, 662, 664, 665, 666, 667, 669, 670, 674, 677, 678, and 686.
United Nations Security Council Resolution 707, S/RES/707, 15 August 1991, adopted by the Security Council at its 3,004th meeting, on 15 August 1991.
'Unlisted Pathogens Relevant to the Convention', Working Paper Submitted by Cuba, Seventh Session, BWC/AD HOC GROUP/WP.183, 22 July 1997.
United Nations Security Council Resolution 715, S/RES/715, 11 October 1991, adopted by the Security Council at its 3,012th meeting, on 11 October 1991.
UNSCOM, United Nations, S/1995/864, 11 October 1995.
United States, *Technical Aspects of Biological Weapon Proliferation,* Office of Technology Assessment (OTA), 1993.

5 Declassified United States documentation

Annual Report of the Chemical Warfare Service, FY 1945, Office of the Chief Chemical Warfare Officer, Washington, DC.
Biological Agents and Munitions Data Book, Miscellaneous Publication 34, Systems Analysis Division, Department of the Army, Fort Detrick, Maryland, November, 1969.
CIA, *Quarterly Review of Biological Warfare Intelligence,* First Quarter, 1949, OSI/SR-1/49.
Colonel William M. Creasey, Chief, Research and Engineering Division, Office of the Chief, Chemical Corps, Report to the Secretary of Defense's Ad Hoc Committee on CEBAR, 24 February 1950.
'Feathers as Carriers of Biological Warfare Agents', Special Report, No. 138, Special Operation and Crops Division, 15 December, 1950. Declassified, 29 August 1977.
Foreign Service Despatch, No. 201, American Consulate General, Hong Kong, 1957.
Joint Chiefs of Staff, 1837/8, 13 September 1949.
Irish, K. *Anti-Crop Agent Munitions Systems,* BWL Technical Study 12, Annex B to Appendix VI, July 1958, of Operations Research Study Group, Study No. 21, U.S. Army Chemical Corps Operations Research Group (ORG) Army Chemical Center, Maryland, Volume 14, p. 4, ORG-0511 (58), AD 347148, 1 August 1958.
Laughlin, L.L. *U.S. Army Activity in the U.S. Biological Warfare Programs,* Vol. I, 24 February 1977.

Laughlin, L.L. *U.S. Army Activity in the U.S. Biological Warfare Programs*, Vol. II, 24 February 1977.
Miller, D.L. History of Air Force Participation in Biological Warfare Program, 1944–1951, Historical Study 194, Wright Patterson Air Force Base, September 1952, Part II.
Miller, D.L. *History of Air Force Participation in Biological Warfare Programme, 1944–1951*, Historical Study No. 303, Wright-Patterson Air Force Base, January 1957.
Military Biology and Biological Warfare Agents, Department of the Army, TM 3-216, 1956.
Minarik, C.E. and Latterell, F.M. *Anti-Crop Agents*, BWL Technical Study 11, Office of the Deputy Commander for Scientific Activities, Fort Detrick, Frederick, Maryland, 1958, Fort Detrick Control Number 58-FDS-666.
Note by Joint Secretaries, *Potentialities of Weapons of War During the Next Ten Years*, Chiefs of Staff Committee, Joint Technical Warfare Committee, 12 November 1945. Enclosed Report by the Inter Services Sub-Committee on Biological Warfare.
'Note by the Secretaries to the Joint Intelligence Committee on Soviet Capabilities for Employing Biological and Chemical Weapons', J.I.G. 297/2, 27 January 1949.
Perkins, W.A., McMullen, R.W., Vaughan, L.M., *Current Status of Anti-Crop Warfare Capability, Appendix VI,* Operational Effectiveness of Biological Warfare, Operations Research Group Study, No. 21, Vol. 13 (of 15 Volumes), 1 August 1958, US Army Chemical Corps Operations Research Group, Army Chemical Center, Maryland, AD 347147.
'Report by the Joint Strategic Plans Committee to the Joint Chiefs of Staff on Chemical, Biological and Radiological Warfare', JCS 1837/18.
'Report on Scientific Intelligence Survey in Japan', Volume V, 'Biological Warfare', September and October 1945.
'Report of the Secretary of Defense's Ad Hoc Committee on Biological Warfare', 11 July 1949.
'Report of the Secretary of Defense's Ad Hoc Committee on Chemical, Biological and Radiological Warfare', 30 June 1950.
'Report on Special BW Operations', October 1948.
'Report of the WBC Committee', 19 February 1942.
Roetzel, T.G. *Biological Agents and Munitions Data Book*, Miscellaneous Publication 34, Department of the Army, Fort Detrick, Frederick, Maryland, Fort Detrick Control Number, 69-FDS-1041, p. 82.
Stubblefield, H.I. *Resume of the Biological Warfare Effort*, 21 March 1958.
Summary History of the Chemical Corps: 25 June 1950–8 September 1951, Historical Office, Office of the Chief Chemical Officer, 30 October 1951, p. 3.
Summary History of the Chemical Corps: 9 September 1951–31 December 1952, Historical Office, Office of the Chief Chemical Officer, February 1953.
Summary of Major Events and Problems, Historical Office, Office of the Chief Chemical Officer, 4 September 1953.
Summary of Major Events and Problems, United States Chemical Corps, Fiscal Year 1954, Chemical Corps Historical Office, September 1954.
Summary of Major Events and Problems, United States Chemical Corps, FY 1955, Historical Office, Office of the Chief Chemical Officer, December 1955.

Summary of Major Events and Problems, United States Chemical Corps, FY 1956, Chemical Corps Historical Office, November 1956.

Summary of Major Events and Problems, United States Army Chemical Corps, FY 1957, US Army Chemical Corps Historical Office, Army Chemical Center, Maryland, October 1957.

Summary of Major Events and Problems, United States Army Chemical Corps, FY 1958, US Army Chemical Corps Historical Office, Army Chemical Center, Maryland, March 1959.

Summary of Major Events and Problems, United States Army Chemical Corps, Fiscal Year 1959, Chemical Corps Historical Office, January 1960.

Tullis, E.C. et al. *The Importance of Rice and the Possible Impact of Antirice Warfare*, Technical Study 5, Office of the Deputy Commander for Scientific Activities, Biological Warfare Laboratories, Fort Detrick, Maryland, March 1958, 58-FDS-302.

6 Declassified United Kingdom documentation

'A Note on Herbicide Research, Crop Committee', 1949, AC10031, at PRO, WO195, 10027.

A Note on Spore Suspension Research, Crop Committee, 1949, AC10372, at PRO, WO195, 10368.

'A Note on the Water Content of Cereals, Crop Committee', 1948, AC10268, at PRO, WO195, 10264.

'A Note on the Use of Aircraft in Agriculture, Crop Committee', 1949, AC10546, at PRO, WO195, 10542.

'Anti Crop Detection and Destruction of Destructive Agents on Crop Leaf Samples', Crop Committee, 1952, 11728, at PRO, WO195, 11725.

'Anti-Crop Aerial Spray Trials', Crop Committee, 1954, AC12929, at PRO, WO195, 12925.

'Analysis of Rice Blast Epidemic Caused by *Piricularia oryzae*', Agricultural Defence Advisory Board, 1954, AC13199, at PRO, WO195, 13195.

'Annual Review, 1955', Agricultural Defence Advisory Committee, Advisory Council on Scientific Research and Technical Development, Ministry of Supply, 14 November 1955 at PRO, WO195, 13473.

Biological Warfare Sub-Committee, 'Report on Biological Warfare', Annex to B.W. (47) 20 (Final), 8th September 1947, pp. 27–28.

Cabinet Defence Committee, 13th Meeting, (D(53), noted in Biological Warfare Report 1951–September 1953, 8 December 1953, BW Sub-Committee, BW (53) 19, at PRO, WO188/666.

'Chemical and Biological Warfare', Note by the Joint Secretaries, Defence Research Policy Committee (DRPC), Ministry of Defence, D.R.P./P (62) 33, 10 May 1962. At PRO, DEFE 10/490, 24144.

'Countermeasures to Anti-Crop and Anti-Animal Warfare', Crop Committee, 1949, AC10742, at PRO, WO195, 10738.

D.C.D.R.D., 'Cereal Rust – Processing, Packaging and Storage of Cereal Rust spores', Agricultural Defence Advisory Committee, Advisory Council on Scientific Research and Technical Development, 30 December 1955, at PRO, WO195/13538.

'Effects of Insecticides and Herbicides on Animals', Crop Committee, 1951, AC11568, at PRO, WO195, 11564.
'Experiments at Camp Detrick on Plant Inhibitors on Sugar Beet Crops, Crop Committee, 1952, AC11737, at PRO, WO195, 11733.
Extract from Quarterly Report of the US Chemical Corps', 1948, Crop Committee, AC10216, at PRO, WO195, 10212.
Extract from Quarterly Report of the US Chemical Corps, 1949, Crop Committee, AC10384, at PRO, WO195, 10380.
Extract from US Chemical Corps Research and Development, Quarterly Progress Report, Crop Committee, 1953, AC12648, at PRO, WO195, 12644.
First Meeting of the Crop Committee, Advisory Council on Scientific Research and Technical Development, Ministry of Supply, 10 August 1948, at PRO, WO195/9959.
Heath, G.D. 'A Note of the Recent Work of the Crops Division at Camp Detrick', Agricultural Defence Advisory Committee, Ministry of Supply, 21 December 1955 at PRO, WO195, A.C. 13538.
'Investigations into Anti-Crop Chemical Warfare', Crop Committee, 1949, AC10276, at PRO, WO195, 10272.
'Investigation into Phytotoxicity', Crop Committee, 1950, AC11091, at PRO, WO195, 11087.
Memorandum by the Chairman, B.W. Sub-Committee on BW Policy, Defence Research Policy Committee, Ministry of Defence, 11 May 1950, D.R.P. (50) 53, see note Ø, D.O. (45) 7th Meeting, Minute 5, at PRO, DEFE 10/ 26, 30087.
Minute Dated 30th May 1950 from the Secretary, Defence Research Policy Committee to the Secretary Chiefs of Staff Committee, Note by the Joint Secretaries, Biological Warfare Policy, Defence Research Policy Committee, Ministry of Defence, 9 June 1950, D.R.P. (50) 76 at PRO, DEFE 10/27, 30087.
Minutes, First Meeting of the Crop Committee, Ministry of Supply, 10 August 1948.
Minutes, Fourth Meeting of the Agricultural Defence Advisory Committee, Ministry of Supply, 1 March 1956, at PRO, WO195, AC13640.
Note by the Chief Scientist, Ministry of Supply, 'Agreements Between US and UK on Allocation of Research and Development Effort', Defence Research Policy Committee, Ministry of Defence, 27 January, 1950, D.R.P. (50) 13, p.1, at PRO, DEFE 10/26, 23747.
Note by the Ministry of Supply, Review of the R. and D. 'Programmes – Chemical and Biological Warfare', Defence Research Policy Committee, Ministry of Defence, 8th October 1954, D.R.P. /P (54) 30, p. 2, at PRO, DEFE 10/33.
Note by the Chief Scientist, Ministry of Supply, Advisory Council on Scientific Research and Technical Development, Committee on Crop Destruction, 21 January 1948, at PRO, WO195/9677.
'Report on 2,4-D', Crop Committee, 1950, AC10855, at PRO, WO195, 10851.
'Report on Cereal Rusts', Agricultural Defence Advisory Board, 1955, AC13378, at PRO, WO195, 13374.
'Report on Clandestine Attacks on Crops and Livestock of the Commonwealth', Agricultural Defence Advisory Board, 1950, AC13154a, at PRO, WO195, 13150.
'Report on Recent Work of the Crop Division at Camp Detrick', Agricultural Defence Advisory Board, 1955, AC13538, at PRO, WO195, 13534.

'Report on the Processing, Packaging, and Storage of Cereal Rust Spores, Camp Detrick, Agricultural Defence Advisory Committee, 1955, at PRO, WO195, 13538.
'Screening of *Piricularia* Isolates for Pathogenicity to Rice', Agricultural Defence Advisory Board, 1955, AC13381, at PRO, WO195, 13376.
'Second Report of the Crop Committee', Ministry of Supply, 3 November 1948, at PRO, WO195, 10080.
'Secret Studies in Drying BW Agents', Extract from Camp Detrick Technical Bulletin No.4, Crop Committee, AC12368, at PRO, WO195, 12364.
'Technical Summary Report', 4th Quarter, Camp Detrick, Crop Committee, 1950, AC 10862, at PRO, WO195, 10858.
US Chemical Corps Report for Fiscal Year 1950, AC10761, at PRO, WO195, 10757.

7 Conference papers

Dando, M.R. 'Technological Change and Future Biological Warfare'' paper presented at a conference on 'Biological Warfare and Disarmament: Problems, Perspectives, Possible Solutions', UNIDIR, Geneva, 5–8 July 1998.
Meselson, M. 'The Problem of Biological Weapons', presentation given at the 1818th Stated Meeting of the American Academy of Arts and Sciences, Cambridge, January 13, 1999. Subsequently presented at Pugwash Meeting No. 246, 11th Workshop of the Pugwash Study Group on the Implementation of the Chemical and Biological Weapons Conventions: Implications of the CWC Implementation for the BTWC Protocol Negotiations Norrdwijk, The Netherlands, 15–16 May 1999.

8 Correspondence and memoranda

Bush, V. Office for Emergency Management, Office of Scientific Research and Development, letter to Sir John Anderson, 15 May 1944.
Correspondence between Milton Leitenberg, Robert W. Kastenmeier, and Harry S. Truman, 1969.
Letter from Sir John Anderson to Vannevar Bush, Director, Office for Emergency Management, Office of Scientific Research and Development (OSRD), 19 April 1944.
Letter from Steve Black, Historian, UNSCOM, 28 November 1995.
Memorandum to accompany letter from V. Bush under date of 15 May 1944, to Sir John Anderson.
Sir John Anderson, Memorandum to the Prime Minister on Crop Destruction, 26 June 1944.
Trevan, T. Formerly Special Advisor the Executive Chairman and Spokesman for the UN Special Commission, January 1992-September 1995, letter to Professor Paul F. Rogers, 10 November 1997.

9 Monographs

Pearson, G.S. 'Article X: Further Building Blocks', Strengthening the Biological Weapons Convention, Briefing Paper No.7, March 1998.
Pearson, G.S. and Sims, N. 'National Implementation Measures', Strengthening the Biological and Toxin Weapons Convention, Briefing Paper No.4, January 1998.

Tansey, G, Trade, 'Intellectual Property, Food and Biodiversity', A Discussion Paper, Quaker Peace and Service, 1999.
Wheelis, M. 'Outbreaks of Disease: Current Disease Reporting', Briefing Paper No.21, April 1999.
United States, 'Technical Aspects of Biological Weapon Proliferation', Office of Technology Assessment, 1993.

Index